T0220484

The Language of Symmetry

The Language of Symmetry is a re-assessment of the structure and reach of symmetry, by an interdisciplinary group of specialists from the arts, humanities, and sciences at Oxford University. It explores, amongst other topics,

- order and chaos in the formation of planetary systems
- entropy and symmetry in physics
- group theory, fractals, and self-similarity
- symmetrical structures in western classical music
- how biological systems harness disorder to create order

This book aims to open up the scope of interdisciplinary work in the study of symmetry and is intended for scholars of any background—whether it be science, arts, or philosophy.

Benedict Rattigan is a writer, and he is the Director of the Schweitzer Institute, a think-tank that promotes an ethic of 'reverence for life' through educational programmes, community outreach projects, a peer-reviewed journal, and university conferences.

Denis Noble held the Burdon Sanderson Chair of Cardiovascular Physiology at Oxford in 1984–2004 and was appointed Professor Emeritus and Co-Director of Computational Physiology. Professor Noble is one of the pioneers of systems biology and developed the first viable mathematical model of the working heart. He is one of the founders of the new field of Systems Biology and is the author of the first popular science book on the subject, *The Music of Life* (2006).

Afiq Hatta is a science writer and the owner of the blog *simplesalad.ghost.io*. He primarily writes about theoretical physics and its connections with economics and philosophy. He is a recent graduate of Mathematics at Trinity Hall, Cambridge, and was formerly a quantitative trader at Morgan Stanley.

The Language of Symmetry

Edited by
Benedict Rattigan, Denis Noble and Afiq Hatta

Illustrations by Alexandra Miron

CRC Press
Taylor & Francis Group ·
Boca Raton London New York

CRC Press is an imprint of the
Taylor & Francis Group, an **informa** business

A CHAPMAN & HALL BOOK

First edition published 2023
by CRC Press
6000 Broken Sound Parkway NW, Suite 300, Boca Raton, FL 33487-2742

and by CRC Press
4 Park Square, Milton Park, Abingdon, Oxon, OX14 4RN
CRC Press is an imprint of Taylor & Francis Group, LLC

Library of Congress Cataloging-in-Publication Data

Names: Rattigan, Benedict, editor. | Noble, Denis, 1936- editor. |
Hatta, Afiq, editor.
Title: The language of symmetry / edited by Benedict Rattigan, Denis Noble CBE, FRS, University of Oxford, UK, Afiq Hatta ; illustrations by Alexandra Miron.
Description: First edition. | Boca Raton, FL : C&H/CRC Press, 2023. | Includes bibliographical references and index.
Identifiers: LCCN 2022041644 (print) | LCCN 2022041645 (ebook) | ISBN 9781032308494 (hbk) | ISBN 9781032303949 (pbk) | ISBN 9781003306986 (ebk)
Subjects: LCSH: Symmetry. | Symmetry (Art) | Symmetry (Physics)
Classification: LCC BH301.S9 .L36 2023 (print) | LCC BH301.S9 (ebook) | DDC 117--dc23/eng/20221107
LC record available at https://lccn.loc.gov/2022041644
LC ebook record available at https://lccn.loc.gov/2022041645

ISBN: 9781032308494 (hbk)
ISBN: 9781032303949 (pbk)
ISBN: 9781003306986 (ebk)

DOI: 10.1201/9781003306986

Typeset in Minion
by Deanta Global Publishing Services, Chennai, India

Contents

Preface *by Professor Denis Noble CBE FRS*

IS OUR WORLD RULED by order or by disorder? It is not just a big question. It is an old question that has engaged philosophers and scientists for thousands of years. We can trace it back to ancient Greek philosophers, to Egypt's pharaoh dynasties, to ancient Chinese Taoists, and to Buddhism's origins in India. And it is also a question that is very relevant to life in the West today.

Towards the end of 2019, before the coronavirus pandemic hit the world, I was visited in my laboratory by the writer and philosopher Benedict Rattigan, who invited me to join a fascinating project he was launching. He presented an idea that we have fundamentally misunderstood symmetry, and that its reach is far more extensive than has previously been recognised.

Rattigan's theory intrigued me, and we worked together to invite other leading Oxford professors specialising in multiple disciplines to explain in lay terms the role of symmetry in their fields of mathematics, music, logic, philosophy, physics, chemistry, astronomy, and biology. Their lectures were delivered at a public event at the British Museum in January 2022, and they form the basis for this collection of essays.

What is the common theme running through the various 'languages' of symmetry? There may not be just one common theme, of course. The concept of symmetry is used in significantly different ways, for example, in physics, where symmetry-breaking plays a central role in explaining the evolution of the universe, and in its use in art, where good proportions are key. But there is one sense of symmetry that stands out from the variety of forms: a fundamental symmetry is that between order and disorder.

It may seem paradoxical to suggest that Nature's *ordering* principle encompasses a symmetry between order and... well precisely disorder. Virtually all 19th-century scientists, and most 20th-century scientists, would have recoiled in horror at such an idea. But what we found as the talks were drafted was that each form of the language of symmetry revealed a new aspect of the order–disorder principle.

In my own field of genetics, for example, I have been able to demonstrate that rather than disorder being passively experienced by living organisms, they harness that randomness and use it as a tool with which they can generate many possible solutions to environmental challenges. I presented a paper on this subject at The Royal Society in 2016, but at the time I didn't see it as a symmetry issue. More recently, however, I have come to understand why this process happens: it places the harnessing of disorder as part of a deeper symmetry process by which order and disorder interrelate.

Empirical evidence for the symmetry between order and disorder is hiding in plain sight all around us. In 1999, the Danish physicist Per Bak suggested to a group of neuroscientists that the brain works on the same fundamental principles as a sand-pile. Imagine an hourglass. Grain after grain, sand falls from the top of the hourglass to the bottom. The pile of sand at the bottom of the hourglass becomes increasingly unstable, and at any moment a single grain of sand might cause a small avalanche. When this happens, the base of the sand-pile widens, which increases its overall stability, after which the process repeats itself. Bak observed that the sand-pile maintains order by means of these random and unpredictable avalanches, which is an example of disorder being 'harnessed' by the sand-pile as a means of maintaining order. In other words, there is a fundamental interdependence between them. It has been demonstrated that this model can be applied to a wide variety of different phenomena, from financial markets and traffic flows to earthquakes, black holes, and the distribution of galaxies in the universe.

Another example of order and disorder forming a symmetry can be found in that most random of phenomena, crowd behaviour. Rattigan writes:

> Watching a group of people negotiating a bottleneck at the doors as they filed into a lecture theatre, I observed what I, as a layman, would describe as chaos. But any social scientist will tell us that the behaviour of a cluster of people conforms to broadly predictable patterns, patterns that are determined by factors such as geometric boundary conditions or time gap distributions. It's these factors which are calibrated in order to improve design elements in lecture halls or theatres. So what I perceive as chaos is, in fact, randomness existing within a framework of order ... Life seems to exist at this border between order and randomness.

As the team at Oxford explored symmetry in their different disciplines, we realised just how multifaceted it is. One is aware of it in musical composition as assonance and dissonance, or you might catch sight of it in biology as a transformation that leaves an object unchanged. Frequently it is elegantly simple, whilst occasionally it can be so contradictory as to be almost incomprehensible. It is both orderly and disorderly, logical and illogical, transparent and opaque. But why is symmetry so symmetrical? The conclusion we reached is this: because it conditions its own structure. Whereas all the laws of Nature are built upon certain symmetries, symmetry alone is subject to itself.

In Nature's laws, complexity never develops beyond necessity. The best designs are simple, and the best designer is Nature itself. Finding that there is a fundamental symmetry between order and disorder that can run through all our explanations is like discovering a clarification that was always waiting there to be revealed. Whereas this seems at first to be illogical, if symmetry were life's ordering principle then it would necessarily have infinite reach and encompass everything, *including itself.* This would create a paradoxical universe of symmetry and asymmetry, of cosmos and chaos. Our universe, in other words, a place of infinite symmetries in which the co-existence of order and disorder is not only in evidence, it is inevitable. If symmetry has no deeper cause than itself, then the cosmos has been structured in the only way possible.

Denis Noble
Balliol College, Oxford

Contributors

The Language of Symmetry is written by an interdisciplinary group of specialists from the arts, humanities, and sciences at Oxford University

Caroline Terquem—Professor of Physics

Dimitra Rigopoulou—Professor of Astrophysics

Alan Barr—Professor of Particle Physics

Joel David Hamkins—Professor of Logic

Sir Anthony Kenny—Department of Philosophy

Robert Quinney—Associate Professor of Music

Anant Parekh—Professor of Physiology

Denis Noble—Emeritus Professor of Cardiovascular Physiology

Edited by Denis Noble, Benedict Rattigan, and Afiq Hatta

Introduction *by Benedict Rattigan*

IN A GAME OF chess, a player must have the ability to visualise all sorts of positions that are reached when certain moves are made, but without actually making them on the board. They should be able to see them so clearly that they can accurately evaluate each new position. The difference between a middle-ranking chess player and a Master can be measured in part by their ability to visualise a wide variety of different outcomes.

Innovation and discovery in science likewise require a degree of creativity. Without imagination, it would be impossible to fill the gaps in our perceptions of the world and develop workable theories for what could be going on beneath the surface.

And yet imagination and science are uncomfortable partners. Whereas speculation pushes outwards without set boundaries, empirical knowledge confines us to what can be observed by the senses and pulls us back in, and this contradiction sits uneasily with many scientists.

But we can entertain a creative thought without necessarily accepting it, and when combined with a rational analysis of any new hypothesis, this openness to new ideas can sometimes yield spectacular results.

The quest for life's ordering principle is the greatest intellectual puzzle of them all. Scientists and philosophers are attempting to identify the law that directs all of Nature's other laws and makes them behave as they do. Solving this conundrum would enable us to answer some fundamental questions about life. It should help explain the origins of the universe, and its discovery could have a profound impact on our understanding of the world and humankind's place in it.

Complex effects arise out of simple causes, and Nature's ordering principle is likely to be elegantly simple. In the words of the Nobel prize-winning physicist Leon Lederman, it should be simple enough to print on a T-shirt. And this elusive principle must be hiding in plain sight all around us, in everything we touch and see, for a law that applies in one place in the universe holds true everywhere the same.

Our understanding of the universe is based upon two different theories. On the one hand we have the model of general relativity theory, which explains to us the large-scale universe of planets and galaxies, and on the other hand we have a completely different set of rules, quantum mechanics, which tells us about the small-scale, subatomic universe. Whereas general relativity presents us with a model of space and time in which things behave smoothly and predictably, the rules that govern quantum mechanics are random and apparently chaotic. Scientists are baffled by this: both models are known to be correct, but they're contradictory. It's as if there are two completely different sets of rules at work.

Modern physics is full of contradictions like this. We find order and chaos, matter and antimatter, and subatomic particles that are both destructible and indestructible. At its most fundamental level, the world of physics is brimming with paradoxes. The ordering principle must be sophisticated enough to explain these contradictions to us.

But why has a law that is likely to be elegantly simple—*a principle that must be all around us*—never been discovered?

Quantum physics has given us a fresh perspective on the world. We have learnt that under certain experimental conditions, light can present itself in contradictory ways. To some observers it takes the form of a particle, which is confined to a small space, or it can appear as a wave, which is spread out. In other words, it can appear to be both 'drawn in' and 'drawn out' at the same time. Light can reveal either one of these aspects, particle or wave, to different observers under different circumstances.

This notion of a particle–wave duality was unsettling to scientists when it was first advanced. It had always been assumed that any law or process of Nature must be logical and objective, something that can be written down and dissected like $E = mc^2$ or Archimedes' principle. But this was different. Unexpectedly, the picture which emerges from the world of subatomic particles is one of contradiction.

Nature's ordering principle is the law which gives rise to these paradoxical and subjective conditions, yet we assume that it must be both logical and objective. But, as with particle and wave, is it not possible that this principle is multifaceted, presenting itself in contradictory ways under different circumstances? Perhaps it is logical and orderly under certain conditions, and at other times it is random and chaotic; it might even express itself as randomness and order, simultaneously. Who says that, at its deepest level, our universe must be comprehensible?

*

In this section, I will take you around some of the artefacts in the British Museum's collection to demonstrate how symmetry was interpreted by different civilisations over the centuries. And, in doing so, we will see that several ancient societies developed a subtly different understanding of this concept from our own, but their interpretation of symmetry makes perfect sense when we apply it to the apparent contradictions of quantum mechanics.

Symmetry is a profoundly important principle. It is central to mathematics, and it has emerged as one of the most fundamental ideas in science, hidden at the heart of everything from quantum mechanics to Einstein's theory of relativity. In fact, it is only slightly overstating the case to say that physics is the study of symmetry. But symmetry is not a new idea. Many different cultures in the ancient world believed it to be life's ordering principle.

This clay bowl was made in Mesopotamia more than 8,000 years ago, and it is as old as civilisation itself. The Mesopotamians observed that the cycles of Nature all follow an identical four-fold pattern. The seasonal cycle, for example, follows a sequence of spring and summer, autumn and winter, and they illustrated this by means of an equilateral cross or a swastika, which is formed here by four goats.

Each phase is counterbalanced by that feature which stands on the opposite side of the cycle—for example, dawn and dusk on the daily cycle, or the ebb and flow of the tide. The bent arms of the swastika emphasise the dynamic character of the cycle, which is forever moving on towards the next phase like a wheel.

As farmers and fishermen, it was important to the peoples of the ancient world that they understood the patterns of nature. What they observed was a series of simple rhythms, and they concluded from these that balance or symmetry is an essential feature of the natural order.

But they also observed a second principle in the world, one which is just as prevalent as symmetry but which seems to be an expression of increasing disorder: people grow old, and objects decompose. All that's born, they observed, must die. Everything around us—plant, mineral, or animal—undergoes a process of eventual degeneration and decay.

The ancient Egyptians understood these two principles as *order* and *chaos*, and they saw their conflict in terms of a vast cosmic battle. Since the dawn of time, the goddess of order, Ma'at, has been at war with Isfet, the god of chaos. And the Egyptians didn't hesitate when deciding where their loyalties lay: they became warriors for Ma'at.

Almost everything the ancient Egyptians did, from the development of their social structures to their religious beliefs, their art, and their architecture, had one specific end in view: maintaining the ascendancy of order over chaos.

Death and decay were seen to be expressions of disorder, and from about 4000 BC, it became customary to mummify the bodies of the dead. Certain words spring to mind as we stand in front of this Egyptian coffin, like '*symmetrical*', or '*orderly*'. On the front, with her wings outstretched, we find the goddess Nut, but in stark contrast to Nut's flowing elegance, the figure itself is static and rigid. The eyes are fixed straight ahead, suggesting control and composure. The craftsmen who made coffins like this were expected to stick to a strict set of rules or guidelines. Their role wasn't to create an object of beauty or to satisfy any wider aesthetic sense, but rather to give form to the inherent symmetry of the universe.

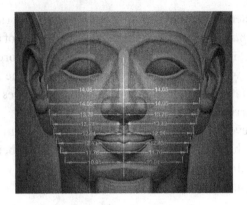

The language of symmetry has many dialects, and the harsh reality of living in an environment which was by turns parched and flooded led the ancient Egyptians to speak of symmetry in an entirely different way from other peoples of the ancient world. Whereas the Mesopotamians saw invariance and harmony as life's ordering principle, the Egyptians believed in a battle between order and chaos in the world. Their divinely appointed task was to impose symmetry on the dangerous and chaotic landscape of life.

Sometime before the 5th century BC, philosophers made an important observation about symmetry. Most people tend to think of symmetry in terms of invariance or harmony, but that's only half the picture.

A definition of balance is 'the presence of two or more equal but opposing principles'. On a set of weighing scales, for example, there are two pans of *equal* weight set on *opposite* sides of a pivot. There can be no balance without two complementary factors: equality and *opposition*. The Ancient Greeks observed that the same principle holds true with symmetry. Without transformation or *variance* of some sort—either over time, in location, or in direction—there cannot be symmetry.

By mixing together the opposing qualities of repetition and variance, the Ancient Greeks came up with some of the earliest examples of aesthetic beauty as it's understood today, and this can be seen in their art and design. With this amphora vase, for example, the different parts of the vase complement each other whilst simultaneously creating contrast They observed that rigid repetition, as favoured by the Egyptians with the pyramids, for example, is often stale and boring. In other words, rather than fight against the discord which is an essential feature of life, they chose instead to celebrate it.

 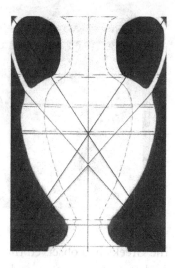

So the Greco-Roman world had a more sophisticated understanding of symmetry than the Egyptians, and this is reflected in their different attitudes towards death. Let's compare this Etruscan funeral urn with the Egyptian coffin we saw a few minutes ago. Rather than being rigidly symmetrical, the figures here seem to be positively laid back. The panel on the front shows a man and a woman bidding each other a fond farewell before departing into the afterlife. Above this, we find the figure of the woman reclining on a couch. You'll see that she holds a good-will gift for the gods on her lap, a bowl containing food or drink. There is no suggestion here of a battle between the forces of order and chaos, and no resistance against the inevitable degradation of death.

The Greeks had observed that some sort of contradiction lies at the heart of symmetry, and the Romans believed that there was something sacred about this idea. Many of their religious symbols comprise contrasting shapes amalgamated within a single harmonious design—for example, circles within squares, or triangles within circles. These symmetrical motifs can be found at some of their most sacrosanct sites and temples.

The Pantheon was constructed almost 2000 years ago, and it's full of symbolic references to this sacred law of the ancient world. The building is wholly symmetrical: left hand balances right hand, and the width is proportionate to the height. Yet within this harmonious framework, there are strong oppositional elements. The monotony is intentionally broken up by a balance of dissimilar proportions, creating a discordant harmony.

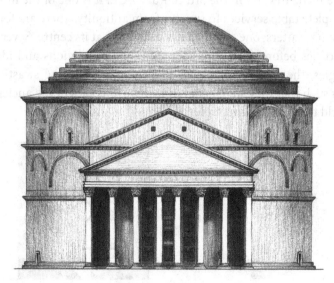

Inside the building, further contradictions reveal themselves. The squared circle was a symbol representing the unification of primary opposites; and the main chamber is topped by a vast dome which forms one half of a perfect sphere, reflecting the symmetry of the universe.

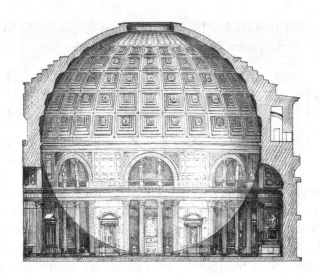

The British Museum also has a collection of silver dishes that were made in Gaul in the 3rd century BCE. It is one of the most complete table services to survive from antiquity. There are four serving platters, one of which has a swastika at its centre. Several centuries before they were adopted by other religions and ideologies, the sanctity of motifs like the cross and the swastika was such that they were believed by many peoples of the ancient world to ward off ill-fortune.

And this symbol (below) is known as the *taijitu*. It represents what the ancient Chinese believed to be the first law of life. Taoist philosophers observed that the cycles of nature all follow an identical polar pattern. With the seasonal cycle, for example, the days grow progressively longer during spring and early summer, whereas after the midsummer solstice they gradually decrease in length. The dots are a reminder that each side of the polarity contains the seed of its opposite, so the growth of spring, for example, automatically generates the promise of decay.

This same two-fold pattern is prevalent throughout nature, conditioning everything from the daily and seasonal cycles to the waxing and waning of the moon. The sheer obviousness of this pattern didn't diminish its significance to the Chinese, for they believed that the universal principle must be in evidence all around us.

Coins or charms such as these were believed to possess special properties, for like the *taijitu* they symbolised the sacred union of opposites.

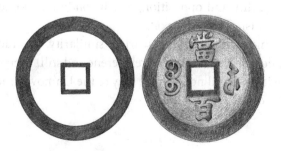

In the West, we tend to get quite confused by the idea of the implicit unity of opposites. It seems to us counter-intuitive that they should be in any way connected. But Taoist sages from ancient China placed particular emphasis upon the oppositional nature of balance; so much so that the notion of polarity came to influence almost every aspect of their beliefs.

Their proof of this ancient Oriental 'theory of everything' lies in an elegantly watertight argument: the first law of life, they reasoned, must govern all the cycles of nature; and nature's cycles share just one principle in common—polarity.

Like other peoples, the philosophers of ancient India encountered both order and chaos in the world. But unlike the Egyptians, who saw the relationship between these two principles in terms of a battle, the early Hindus understood chaos as an aspect of a *deeper symmetry*, a symmetry of order and disorder.

This is Shiva. As Nataraja, the cosmic dancer, he or she is depicted with several arms, each representing a different aspect of life. One hand upholds the drum of order and creation, and the beat of time; and in another hand we find the flame of chaos or destruction. Often the figure wears both male and female earrings, and it's not uncommon to find depictions of Shiva in which they're androgynous, for he—or she—unites and transcends all opposites.

Lower down, one leg is raised and Shiva is stamping their heel to the rhythm of the universal drum. The drumbeat represents the concept of discordant harmony, or the simultaneous presence of equality and opposition, which conditions not just balance, but musical rhythm too.

And amidst this dizzying swirl of similarity, contradiction, and paradox, Shiva's head is entirely serene and still. Their expression is neither joyful nor sad; for they're the Unmoved Mover, at the centre of the circle of life.

So a belief in a universal law of symmetry extended from ancient Egypt, China, and India to Greece and Imperial Rome. Most of these peoples came up with this idea entirely independently of each other. And whilst they placed emphasis upon different aspects of symmetry—in Egypt, there was believed to be an irreconcilable conflict between order and chaos, for example, whilst Hindu scholars saw a mutual dependence or a *synthesis* between these two principles—all of them held symmetry in such high regard that it conditioned everything from their art and architecture to their political structures, the layout of their cities, and their religious beliefs.

We've seen that symmetry is a central paradigm of physics, but a law which applies in one place in the universe holds true everywhere, and life's ordering principle—if such a law exists—must be hiding in plain sight all around us. Let us take the British Museum's café area as an example. There's not much evidence of an ordering principle here! In a busy cafe teeming with people, everything's chaotic. But let me show you something.

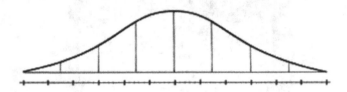

This rather elegant and smooth design is known as a Gaussian curve. It's widely used by statisticians to evaluate probabilities in apparently random behaviour. If you throw a set of dice, for example, there's a 100% probability that the value on the face of the dice won't exceed 12, and there is a 0% probability that the outcome will be less than 2. But with a great many throws of the dice, the most common outcome will be the number 7, which is mid-way between 2 and 12, and the normal distribution of numbers will polarise around this central number. Perhaps God plays dice with the universe—and nothing's more random than the throw of a set of dice—but even this can be seen to be subject to an underlying symmetry.

So how does this relate to a group of people in a café? Well, if we consider factors such as gender, ethnicity, and cohesiveness, all of these fall within the confines of a Gaussian curve. What we're seeing here, in fact, is randomness and order coexisting as parts of a deeper symmetry—a Dance of Shiva! And this principle is in evidence all around us, although it expresses itself in a variety of ways. In fact, many manifestations of symmetry are so commonplace that we take them for granted as everyday facts of life, just as you'd expect of life's ordering principle. The 18th-century economist Adam Smith identified what he called a 'hidden hand' controlling the process of supply and demand, for example: if a shopkeeper undercharges for their goods they'll be unable to meet their overheads, but if they charge too much they'll be undercut, so what they seek is the mid-point between opposites.

COFFEE

Espresso	1.85	CAPPUCCINO	2.75	3.00
MACCHIATO	1.85	LATTE	2.75	3.00
RISTRETTO	1.85	MOCHA	2.90	3.20
ESPRESSO CON PANNA	2.20	AMERICANO	2.45	2.75
CORTADO	2.45	FLAT WHITE		2.80

TEA ENGLISH BREAKFAST HERBALS & INFUSIONS	2.10	HOT CHOCOLATE WITH WHIPPED CREAM	2.90	3.20

PREFER A MILDER COFFEE? JUST ASK FOR ONE ESPRESSO SHOT IN YOUR REGULAR OR GRANDE COFFEE

This gets far more complex when other factors are involved, of course, and we'll consider this later; but symmetry likewise lies at the heart of this complexity.

And this same principle, symmetry, can be seen here on a menu board. Most cafés offer a wide variety of different drinks to choose from—cappuccinos, frappuccinos, and so on. But studies suggest that too many options can lead to a sort of analysis paralysis. This process is known by retailers as 'the paradox of choice': whilst consumers like to be able to choose, too many options can be off-putting.

And in fact paradox is a lot more commonplace than people realise. For example, a shop or café is a social system, and social scientists have observed that all such systems have two contradictory tendencies. On the one hand, they tend towards chaos—a significant part of a social organisation's energy is devoted to overcoming entropy and maintaining structure. But social structures also have a tendency to return to their original order when they're disturbed. It's known as 'social equilibrium': change in one element is frequently followed by corresponding changes in related elements, which work towards diminishing the original disturbance.

Symmetry—it's hiding in plain sight all around us. We should reawaken to what we once knew: that it encompasses order and chaos, harmony and opposition, and logic and paradox. This ancient understanding of symmetry is as old as civilisation itself, and I believe the time has now come for a re-evaluation and a refinement of ages-old wisdom.

Benedict Rattigan
The British Museum

Editors' Note
by Afiq Hatta

*T*HE LANGUAGE OF SYMMETRY is a collection of essays that highlight the role of symmetry and asymmetry across a wide range of academic disciplines including physics, philosophy, biology, and music. While some of the essays focus exclusively on particular facets of symmetry or asymmetry individually, one significant aim of this book is to introduce examples of a more profound correspondence that exists between symmetry and asymmetry.

It is not surprising that the topics of order and disorder appear as central themes in many academic fields. An informed reader might observe that these topics are already well established in mathematics and physics. For example, the essay by Caroline Terquem eloquently highlights how ordered relationships in planet orbits emerge from chaotic and disordered discs of gas. This emergence of order is not only an explicit but a rather beautiful example of the interplay between order and disorder.

There are also fascinating examples of the interplay between randomness and order in biology. As Denis Noble writes, there is overwhelming evidence that living organisms harness disorder and stochasticity with purpose—a theory that challenges the neo-Darwinist concept of 'blind stochasticity'. In biology, systems have been developed to defend against randomness. For example, neurons need to figure out how to reduce signal noise when transmitting electrical impulses so that our brains do not get confused. However, there is evidence that organisms harness randomness to generate new functions to better survive against a change in their environment in challenging scenarios. In his essay, Noble argues that there are cases when randomness is used with purpose rather than just passively. In evolutionary biology, function is why some object or process occurred in a system that evolved through natural selection. In physiology, function is a purpose. Noble references studies that organisms use randomness to generate responses to high environmental stress. When an organism encounters a challenge, it relies on randomness to try and find something that works. Noble provides two examples. Firstly, in the immune system, the mutation rate in the variable part of the genome is accelerated in response to a new antigen. Secondly, in bacteria, there is a five-order of magnitude acceleration of reorganisations in the genome to help cope with additional environmental stress.

Within academic fields, order and disorder have traditionally been studied separately. In physics, for example, order and symmetry have related yet subtly different meanings. The word 'order' is usually associated with a system in a state of low entropy. Entropy, in this context, describes how *mixed-up* the possible states in a physical system are. To understand the concept of entropy better, let us do a little thought experiment. Imagine closing your eyes and then flipping a single, fair coin. When you open your eyes, the coin can occupy one of two states. However, imagine a system of 10 coins. In this case, we have 1,024 different configurations and, therefore, many more possible outcomes. A system with low entropy is a phrase used to describe a system with relatively few, concentrated outcomes across a probability distribution.

The study of symmetry in physics has been well established since the work of Emmy Noether was published in 1918. In this context, symmetry refers to transformations of physical systems that leave the systems unchanged. For example, imagine a square without labels. If you were to rotate a square 90 degrees, it would be unaffected. Similarly, it would remain unchanged if you were to reflect it across one of its axes. A transformation might be as simple as rotating a system through a certain angle. On the other extreme, a transformation could be something as complex as scale invariance in a fractal or the automorphisms that Joel David Hamkins describes in his essay on self-similarity. Noether's theorem states that symmetries which are smooth lead to conserved quantity laws, like the conservation of energy. The analysis of a physical theory starts with the Lagrangian, which is a system's kinetic energy subtracted by its potential energy. The construction of a Lagrangian is usually motivated by symmetries, and their shape needs to reflect this in nature. The most apparent set of symmetries a physical theory needs to obey is the Poincaré group. The Poincaré group is one the most natural set of symmetry transformations in nature—it consists of translations, rotations, and boosts that are compatible with special relativity. Symmetry groups that can be parametrised by smooth shapes also have a crucial role in physics and they are called Lie groups.

Dimitra Rigopoulou, in her essay on symmetry in the universe, unpacks the standard model and the concept of entropy. In particular, Rigopoulou considers symmetries in the standard model of particle physics. The standard model of particle physics is a Lagrangian that gives rise to the fundamental particles we have currently observed. However, this model, too, has symmetries, of which it is unclear whether their combinations are genuinely symmetrical. Rigopoulou discusses the three essential symmetries in the concept of the standard model, which are charge, parity, and time reversal.

The study of order is not exclusive to physics, of course. In aesthetics, for example, the word 'order' is intimately related to the concepts of both structure and proportion. In biology, order might refer to the organised structures in living organisms, such as DNA. If we consider the word order in a more relaxed light, we could say that the definition of order might include symmetry.

While subtly different from the word 'order', 'symmetry' plays a significant role in these fields.

On the flip side, the word 'disorder' suggests a state of randomness. Whilst we will more fully define what chaos and order mean in the following paragraphs, one can still speak about them informally. Both words allude to the study of complex systems. The word 'chaos' suggests a degree of unpredictability. Both physics and the social sciences have led to a very successful enquiry into the study of chaos and disorder. Intuitively, one might think that the meaning of these terms is subtly different, and precise scientific definitions of the words 'chaotic' and 'disordered' are hard to come by. Whilst many philosophers are exploring precisely this question, there are a few principles that one can go by. Firstly, how *stable* is the system? By stable, we mean the sensitivity of a physical system's state to the initial conditions in which it is placed. Physical systems are governed by a system of equations that we have to solve. But, for the most part, these equations only lead us to a 'general shape' of the solution unless we plug in some data about the initial conditions themselves.

For example, imagine a pool table on the first hit. As physicists, we would of course need to know the initial force and direction of the pool cue in order to predict the final state. We then ask ourselves if small changes in the initial conditions lead to significant changes in the final outcome. In the case of a pool game, the intuitive answer is yes—had we nudged the first player to miss her initial shot slightly, we would have ended up with a completely different game. In the same vein, if a physical system exhibits a high degree of sensitivity to its initial conditions, we might consider it a complex system.

The Interplay Between Order and Disorder

Order and disorder, at first glance, are entirely disparate fields, and as such they are usually treated separately. Whilst there is a wealth of literature on these concepts individually, there is very little published material that links order and disorder together.

However, there is strong evidence emerging to suggest an underlying correspondence, or mechanism, between symmetry and asymmetry itself. This correspondence may be described as *a symmetry of symmetry and asymmetry*. There are some explicit examples of this rather curious symmetry in physics. For example, as Professor Rigopoulou explains, symmetry-breaking occurs at the transition when a physical system with a symmetric Lagrangian enters a ground state that is no longer symmetric.

Of course, conserved quantities are not the only use of symmetries in modern physics. One surprising concept is the role of symmetry *in creating particles themselves*. This process is called spontaneous symmetry-breaking, and it is currently the best mechanism we have to describe why particles have mass. A system might break from a symmetric state into a non-symmetric one at low energies. When this happens, we find that weird ephemeral particles pop into existence near this low-energy state! Whilst this mechanism is used in high-energy theoretical physics, it is widely applicable to something as familiar as a magnet.

This book is a collection of new articles that highlight the role of symmetry, order, and chaos in different academic fields. Throughout, we will also be expanding on the crucial ideas that involve the interplay between order and disorder. The precise definition of 'interplay' is something that will be made clear as we review examples from a range of academic fields.

Afiq Hatta

About the Contributors

Alan Barr is Professor of Particle Physics at Oxford. He has been involved in the ATLAS experiment at the Large Hadron Collider (LHC) since 1999 when that experiment was still in the R&D phase. Part of his research has been development, assembly, and operation of precision semiconductor strip detectors. A separate thread of his research has been searching for signatures of new particles—particularly Higgs bosons and dark matter particles. As well as finding new particles, he has developed many of the techniques which are used to measure their properties—such as their mass, angular momentum, and decay modes.

Joel David Hamkins has made contributions to mathematical and philosophical logic, particularly set theory and the philosophy of set theory, to computability theory, and to group theory. He has held various faculty or visiting fellow positions at the City University of New York, the University of California at Berkeley, Kobe University, Carnegie Mellon University, the University of Münster, Georgia State University, the University of Amsterdam, the Fields Institute, New York University, and the Isaac Newton Institute. In 2018, Hamkins was appointed Professor of Logic in the Faculty of Philosophy at Oxford and Sir Peter Strawson Fellow in Philosophy at University College, Oxford.

Afiq Hatta is a science writer and the owner of the blog *simple-salad.ghost.io*. He primarily writes about theoretical physics and its connections with economics and philosophy. He is a recent graduate of Mathematics at Trinity Hall, Cambridge, and was formerly a quantitative trader at Morgan Stanley.

Sir Anthony Kenny, FBA, is a philosopher. He was Master of Balliol College Oxford (1978–1989) and subsequently an Honorary Fellow. Between 1989 and 1999, he was Warden of Rhodes House (manager of the Rhodes Scholarship program) and he was Pro-Vice-Chancellor of the University of Oxford between 1984 and 2001.

Denis Noble, CBE, FRS, held the Burdon Sanderson Chair of Cardiovascular Physiology at Oxford 1984–2004 and was appointed Professor Emeritus and Co-Director of Computational Physiology. Professor Noble is one of the pioneers of systems biology and developed the first viable mathematical model of the working heart. He is one of the founders of the new field of Systems Biology and is the author of the first popular science book on the subject, *The Music of Life* (2006).

Anant B. Parekh, FRS, is Professor of Physiology and a Fellow of Merton College, Oxford. He investigates how cells communicate with one another, with an emphasis on how the ubiquitous intra-cellular signalling messenger calcium controls biological functions such as secretion, energy production, and gene expression. Parekh was elected a Fellow of the Royal Society (FRS) in 2019. He was also an elected member of Academia Europaea (MAE) in 2002 and a Fellow of the Academy of Medical Sciences (FMedSci) in 2012. He was awarded the George Lindor Brown Prize Lecture by the Physiological Society in 2012.

Robert Quinney is Associate Professor of Music in the Faculty of Music and a Tutor in Music at New College, Oxford. He is also Organist of New College, directing the famous Choir in daily services, concert performances, and recordings. Quinney began work at Oxford in 2014, following appointments as Sub-Organist at Westminster Abbey and Director of Music at Peterborough Cathedral. In addition to his work at New College, he maintains a busy freelance career as a solo organist and conductor.

Benedict Rattigan is a writer and he is the Director of the Schweitzer Institute, a think-tank that promotes an ethic of 'reverence for life' through educational programmes, community outreach projects, a peer-reviewed journal, and university conferences.

Dimitra Rigopoulou is Professor of Astrophysics at Oxford, Hasselblad Professor at Chalmers University of Technology in Sweden, and a Senior Nicholas Kurti Fellow at Brasenose College Oxford. Prior to her arrival at Oxford, she worked at the Max Planck Institute for Extraterrestrial Physics (MPE) in Garching, Germany (1995–2003). Her research activity focuses on the physics behind the processes that govern star formation in galaxies. She has a long-term interest in the study of the properties of ultraluminous infrared galaxies which are amongst the most luminous objects in our Local Universe and serve as a testbed for understanding galaxy formation and evolution. She has a keen interest in astronomical instrumentation and has been involved in the design of many European Space Agency (ESA) and NASA past and upcoming infrared space missions. She is currently a member of the science team for HARMONI one of the first light instruments on the Extremely Large Telescope (ELT) in Chile.

Caroline Terquem is Professor of Physics at the Rudolf Peierls Centre for Theoretical Physics and a Fellow at University College, Oxford. Her research activity is in the area of the dynamics of extrasolar planetary systems and protoplanetary discs. In the last few years, her research has focused more particularly on the gravitational interaction between planets and the disc in which they form, which leads to the so-called migration of the planets toward their host star, thus explaining the existence of planets with very short periods.

CHAPTER 1

Planetary Systems: From Symmetry to Chaos

Caroline Terquem

EDITORS' PREFACE

We will start our journey by discussing the relationships between order and chaos in physics. This section aims to provide technical detail into what these concepts are from a mathematical perspective. Then, we will see these ideas in action through an essay from one of our contributing writers, Caroline Terquem. Caroline is a Professor of Physics at the Rudolf Peierls Centre for Theoretical Physics and a Fellow at University College, Oxford. Her research has focused on the dynamics of extrasolar planets, particularly on the gravitational interaction between planets and the discs in which they form. Her essay examines the intermingling of order and chaos through the lens of planetary motion. In particular, Terquem observes how the seemingly chaotic process of planetary formation yields synchronised systems in their orbits.

DOI: 10.1201/9781003306986-1

The words *chaos, disorder, entropy,* and *stochasticity*—whilst all related—tend to mean subtly different things. We shall start by giving them definitions in the context of modern physics.

Chaotic systems possess the property that they have a high degree of sensitivity to the system's initial conditions. In this context, sensitivity refers to how a system can behave differently even with a slight change to its initial settings. An example of sensitivity would be the evolution of the balls on a billiard table after the initial break. Consider how billiard balls are configured at the end of this initial hit. Over time, you would expect that a small change in the trajectory of the initial hit causes wildly different outcomes in the configuration of the balls, such that it almost 'appears' random.

The definition of chaos is better understood through a concept called phase space, which encodes all the possible states of a system through its physical attributes—usually its velocity and position. For example, consider a ball moving about a one-dimensional line. What are the only two quantities that describe its state of being? The first would be its position relative to some reference point on this line. The second would be its velocity. Together, position and velocity are all that we need to know to infer where the ball will be next. These two quantities together represent a point in the phase space of the ball. As we see this system evolve, the ball will trace out a line in its phase space according to some dynamics.

We can generalise this one-dimensional example of phase space to any sort of physical system we want. In particular, the concept of phase space can be easily generalised to three-dimensional space and can describe multiple objects. Chaos happens when two initially close points in phase space grow apart at exponential speed. This means that if two identical systems start at close points in phase space, we expect their distance from each other to diverge significantly as time passes.

In mechanics, systems that are described as chaotic are not random or stochastic. Most of the time, chaotic systems are adequately modelled with Newtonian physics. Newtonian physics is deterministic. What makes a system chaotic is that in all practical aspects, it is unpredictable due to the initial measurements never being precise enough to inform us of what is going to happen next. We can build complex systems from toy models of simple objects that interact, producing chaotic phenomena that are infeasible to predict. Complex systems are all around us. For example, take the air we breathe. If we imagine each air particle as a single particle, then it is not so hard to model the motion of these things individually on their own. Only when we combine them do we start to see some interesting phenomena.

Conversely, the property of entropy, or disorder, is a story of possibilities. A disordered system should have a lot of opportunities to manifest many different outcomes (like the up-down spins of atoms or electrons on a lattice), with a broadly equal probability of realising those outcomes. We will talk more about entropy in the next chapter.

There are several examples of correspondence between chaos and order in physics—a 'smoothing' out of complexity as one might put it. Professor Terquem's essay focuses on a stunning example of mechanical phenomena where a highly complex, chaotic system tends to become an ordered system through natural means. At the end of the piece, we will see how these ordered systems can reverse back into a chaotic system under the right conditions.

Terquem's essay studies the systems of planets that have a 'clean' ratio of time of orbit. In particular, the periods of the orbit of planets have ratios that are small integers. For example, it takes Dione approximately twice as long to orbit Saturn as Enceladus. In this case, the ratio would be 2:1. There are plenty of other systems that have been discovered to have this sort of mean motion resonance. This is remarkable because one wouldn't expect a violent and chaotic process like planet formation to lead to such a finely-tuned result. The transition is an excellent example of how a disordered physical system can generally tend to a state of orderedness.

'Planet' comes from the Greek word which means 'wanderer', as planets were seen to move relative to the stars. Although the orbits of the planets seemed chaotic viewed from the Earth, the Greek astronomers were convinced that they could be explained by superposing spheres and circles. According to Plato, these were the preferred figures of the natural world. From the geocentric model of Ptolemy to the heliocentric construction of Copernicus, the underlying assumption was that the planets were moving along superpositions of perfect circles. In his *De Revolutionibus*, Copernicus alludes to the symmetry of the Universe which is revealed by the arrangement of these orbs. 'Symmetry', in that context, is taken to mean well-proportioned.

Modern observations of the solar system have confirmed that it is indeed highly structured. The eight planets are well ordered, with the terrestrial rocky planets closer to the Sun, and the more massive (gas or ice) giant planets further away. Each of the giant planets has a cortège of moons, which makes them look like a reduced-scale version of the solar system. These observations are well explained by the theory of planet formation which, in its modern form, originates in the 19th century from the idea that the Earth is made of meteoritic material. This theory was put on a quantitative basis in the second part of the 20th century and has been very successful at explaining the overall order of the planets in the solar system. Planets form in a so-called proto-planetary disc, made of gas and solid particles, which surrounds every newly formed star. The solid particles collide with each other and agglomerate to form bigger and bigger objects, until rocky planets form. Far enough from the Sun, where low temperatures enable ice to condensate, there is enough solid material to produce yet more massive objects, onto which a large amount of gas can be captured by gravity, resulting in giant planets. It is likely that in the same way that planets have formed around the Sun, satellites (moons) have assembled around giant planets.

In the solar system, satellite systems display interesting dynamical effects which are not observed among the planets. Indeed, a significant number of satellite pairs are found to be in *mean motion resonances*, which means that the periods of revolution of the two satellites are commensurable: their ratio is that of two relatively small integer numbers. For example, Enceladus and Dione, two satellites of Saturn, are in a 2:1 mean motion resonance, such that Enceladus, which is closer to the planet, completes two revolutions while Dione completes one revolution. Such resonances may even be seen in systems of three satellites. A famous example, which was first studied by Laplace in the late 18th century, involves Jupiter's moons Io, Europa, and Ganymede, which periods of revolution are in the ratio 4:2:1. Such resonances lead to repetitive configurations of the system of moons, and hence to an enhancement of their mutual gravitational interaction which builds up over time.

How can such a finely tuned and well-synchronised dance be achieved? The naive expectation would be that the process by which moons form leads to a random distribution of periods of revolution. Therefore, moons must subsequently move with respect to each other to establish resonances. This process is subtle, and not understood until the 1960s.

In a seminal paper published in 1965, it was first proposed by Peter Goldreich that moons can be brought into resonance by tidal interaction with the planet around which they revolve. For example, in the case of Enceladus and Dione, tides raised by Enceladus in Saturn result in Enceladus moving away from the planet, in exactly the same way as our Moon recedes from Earth as a result of the tides it raises on our planet. As Dione is further away, it is much less affected by the tidal forces, as these decrease dramatically with distance. Enceladus then migrates closer to Dione, so that the ratio of the orbital period of Enceladus to that of Dione increases. If this ratio is initially smaller than one-half, then it may reach this value at some point. This corresponds to a resonance being encountered. Once the resonance is reached, the enhanced mutual gravitational interaction between the two moons ensures that commensurability is preserved, even as the tidal forces with Saturn keep pushing Enceladus away. Both Enceladus and Dione then remain locked in resonance as they both migrate away from the planet. The same process may also explain how Io, Europa, and Ganymede are in the so-called Laplace resonance described above: the tidal forces of Jupiter push Io away until it "captures" Europa in a resonance. The pair then continues to migrate away as a unit until Ganymede gets caught, at which point the three planets subsequently stay locked in this configuration. A whole system of satellites can be captured that way! This remarkable arrangement would certainly have qualified to be part of the symmetry of the Universe dear to Copernicus. However, as will be shown below, the very same process that can lead to the stable configuration described above can also result in disruption and chaos.

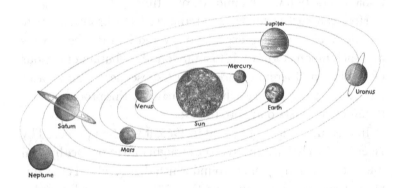

The idea that our own planetary system is not unique in the Universe dates back at least to antiquity. If Nature does not produce truly unique objects, there must indeed be planets around stars similar to our Sun. As stars and planets form together by a mechanism which repeats itself, planets were expected to be found around solar-type stars. The first discovery was not made until 1995 … and it came as a surprise! This is because the planet, called 51 Pegasi b, is a giant similar to Jupiter, but located ten times closer to its host star than Mercury is to the Sun! It is so close to the star that its period of revolution is less than five days. Since then, many more such 'hot Jupiters' have been detected. On the basis of the theory of planet formation mentioned above, only smaller, rocky planets would be expected to be found so close to the star. This is not to say that the theory is wrong. Rather, processes which have not played an obvious role in the context of the solar system have been much more significant for at least some extrasolar planetary systems. These processes are responsible for the so-called *migration* of planets in the protoplanetary disc in which they form. Once planetary cores reach a mass large enough, their gravitational interaction with the gas in which they are embedded changes their orbit in such a way that they may move closer to the star.

This is a process which, in many ways, is similar to the tidal interaction between the Moon and the Earth. As the planetary cores accrete gas and become giant planets, the details of their gravitational interaction with the disc change, but it may still lead to inwards migration. This process, which can explain why some planets are found very close to their host star, had previously been studied in the context of the solar system already in the late 1970s, as it also applies to moons embedded in the rings which surround the giant planets. It had even been predicted that some extrasolar planets could have migrated significantly that way! However, given that the planets in the solar system do not appear to have been subject to significant migration, this prediction had not attracted much scrutiny from observers.

It has been 25 years since the discovery of 51 Pegasi b. Thousands of planets have now been detected, with masses ranging from a fraction of an Earth mass to several Jupiter masses. In almost all cases, planets are not being observed directly. Instead, their presence is inferred from the effect they have on their host star: their gravitational pull results in a wobble of the star that can be detected, or their transit blocks some of the light we receive from the star. Numerous multiple systems, containing up to seven planets, have been observed. In some systems, the presence of a planet is inferred not from its effect on the star, but from how it perturbs the motion of another planet which has already been detected. This is very similar to the way that Neptune was discovered historically, from its measurable perturbations of

Uranus' orbit. It is clear that a lot can be learnt from the disruption of the symmetry of an orbit!

Since planets migrate in their protoplanetary discs and often exist in multiple planet systems, it is not surprising that there are many examples of mean motion resonances. The migration rate depends on many parameters, such as the mass of the planets and their distance from the star. Therefore, situations where the distance between two planets decreases with time are frequent, and this leads to capture into resonances, just as described above in the context of planetary moons in our solar system. A large number of resonances (or near-resonances) involving two planets have been reported, but *chains* of planets in resonances are also common, with up to seven planets involved, such as the Trappist-1 system. These systems are a beautiful illustration of how symmetry and order may emerge from a rather chaotic planet formation scenario.

But chaos may also emerge from order …

The systems we have mentioned above and which exhibit mean motion resonances are stable, which means that there can persist for billions of years without being disrupted by perturbations. However, in some cases, the gravitational interaction between two objects in a resonance can actually lead to the disruption of the system! Evidence of this process is found in the structure of the asteroid belt, which is populated by small rocky bodies and located between the orbits of Mars and Jupiter. These objects, whose sizes are at most a few percent of that of the Earth, are the leftovers of planet formation, which were prevented from growing further because of their gravitational interaction with Jupiter. The distribution of asteroids has voids at locations in the belt corresponding to commensurabilities with Jupiter: these so-called *Kirkwood gaps* (discovered by Kirkwood in the late 19th century) indicate that such commensurabilities are unstable and that any asteroid which might have been present at that location has been ejected. By contrast, there is a clump of asteroids, called the Hilda family, at the location of the 3:2 resonance with Jupiter, located between the main asteroid belt and Jupiter's orbit, which indicates that this particular resonance is stable.

In general, a configuration is more likely to be stable when the asteroid has its closest approach with Jupiter (which happens when all three bodies are in conjunction, i.e. in a straight line) at the asteroid's *perihelion*. This configuration maximises the separation at the closest approach to Jupiter. The interaction between the two objects is therefore rather small in conjunction and does not build up to large values as the geometrical configuration is necessarily repeated (because of the resonant orbits). The asteroids of the Hilda family are actually in this configuration. It was originally suggested by Kirkwood himself that unstable configurations were the results of conjunctions arising when the asteroid was at *aphelion*, which minimises the separation at the closest approach to Jupiter. In this case, the stronger interaction with Jupiter builds up over time and reaches large values. This leads to high eccentricities and/or inclinations causing the asteroids to collide with nearby objects. While this type of instability may indeed happen, more recent advances in celestial mechanics have shown that the trajectories of asteroids in some specific resonances (3:1 for example) are *inherently* chaotic. This is a fundamental property of the equations that govern the motion of the bodies, rather than being caused by some specific configuration of the system. If both orbits were circular, there would be no resonant effect. If only the orbit of the asteroid were eccentric, the resonant gravitational force between the two objects would appear as a single dominant term in the equations, acting at a precise location for the asteroid. Such a configuration may be stable along the lines discussed above. However, even though the eccentricity of Jupiter's orbit is small, it does not vanish, and with both orbits being eccentric, there are in fact several dominant resonant terms in the gravitational force, each giving a strong contribution at their own different locations. Although those locations are distinct, they are close enough to each other in space that they all affect the motion of the asteroid. This is called *resonance overlap*. Recall that one single resonance leads to repeating configurations. The presence of additional resonances, however, produces *uncorrelated* forces which make the motion not repeated, but chaotic. Similar effects happen when distinct mean motion resonances overlap with each other, which is the case in the region between the orbits of the asteroids of the Hilda family and Jupiter's orbit. For example, the 4:3 resonance overlaps with the 5:4 resonance, which itself overlaps with the 6:5 resonance (the ratios being close to each other), and all resonances then contribute simultaneously to the motion of the object. This explains why there are no asteroids in this region of space. When resonances overlap, the motion of the asteroid is unpredictable, depending very sensitively on the initial conditions. This is called chaos.

The physics of mean motion resonances in the solar system and elsewhere is rich and fascinating and illustrates how processes that can lead to such elegant configurations seen among the moons of giant planets or ensembles of extrasolar planets can also lead to chaos. While a single resonance acts as a Master Clock in the Universe, the superposition of two resonances yields disruption and unpredictability.

EDITORS' COMMENTARY

Caroline Terquem's penetrating essay shifts our focus onto the meaning of the word 'chaos'—when the resonances of two planets overlap. She observes that the trajectories of asteroids are inherently chaotic. At this point, we would like to once again draw some attention to the differences between chaos and randomness. This is called chaos, but it is not indeterminacy—at every step, we could have predicted the next step if we had infinitely accurate machines. Even determinate equations can give rise to chaotic outputs, meaning we cannot see the order but there is, in principle, an order. Another word for this kind of behaviour would be 'complexity'.

Prior to reading Terquem's essay about this counterintuitive phenomenon, it would seem reasonable to think that chaotic behaviour is likely to occur in systems with many degrees of freedom. However, this is not necessarily the case. This kind of effect is also seen in even the simplest of structures, such as the 'double pendulum'. A double pendulum is a simple structure—it is a pendulum that is freely attached to the end of another fixed pendulum. It can be shown that this structure is also chaotic, according to our definitions above.

There are also other examples of a slightly different nature, where complexity gets washed out in scale, allowing humans like us to model physical systems in the first place. In quantum field theory, the study of renormalisation helps us understand why it is possible to do physics at all. For example, the Casimir effect is a force between two plates in a vacuum. Since atoms have a finite wavelength, we shouldn't need to model the atoms' high energy, ultraviolet interactions with the plates. Instead of putting in the detailed physics of the plates, it is easier to employ practical approximations that wash out the short-wavelength physics that we do not care about. To do this, we can make different assumptions about which interactions and wavelengths to 'wash out' for a finite result. Amazingly, however, it turns out that the force is entirely independent of how we chose to do this procedure. We have experimentally verified that this force is independent of how we model the small-scale physics.

The second approach would be to simplify complex phenomena by modelling them as a random (but well-understood) distribution. For example, it isn't much of a stretch to imagine that the factors that govern someone's height are sensitive to a multitude of factors. However, the large N limit is adequately described by a Gaussian distribution. This distribution is also the case in the laws of thermodynamics—detailed computations of the micro-states get washed out in favour of smooth distributions that describe average relationships between temperatures.

CHAPTER 2

Entropy and Symmetry in the Universe

Dimitra Rigopoulou

EDITORS' PREFACE

In the previous chapter, we examined the interplay between order and chaos through the lens of planetary dynamics. We also find a correspondence between entropy and symmetry in physics. In this context, the word 'symmetry' primarily relates to when a system's dynamics remain unchanged under a transformation. This transformation could be a rotation, translation, or something similar. The type of transformation can be unique at a point, which we call local symmetries, or can affect a system globally.

Entropy, on the other hand, is often associated with the word disorder or chaos. Whilst there are several definitions in physics, one intuitive way we can discuss entropy is by counting the number of states that a system can be in. Consider a system that can be in many different states. In the Gibbs definition of entropy, a system which has the probability of these more uniformly distributed

DOI: 10.1201/9781003306986-2

states is in a higher entropy. In the Boltzmann formulation, assuming that all states are equally likely, the entropy of a system is the Boltzmann constant multiplied by the logarithm of the number of potential states. For example, a coin can occupy two states, so its entropy is k log 2. A collection of two coins can occupy four states, so its entropy is k log 4, which is higher than a system of two coins.

The interpretation of symmetry we will discuss is slightly different from the notion of 'order' in our previous example of synchronised planets. This concept of symmetry is pervasive throughout many academic fields, and as such, it is slightly more difficult to define fully. However, you have probably 'experienced' the concept of symmetry through different ways. For example, we can stare at a mirror and see ourselves, which is symmetry in the form of reflections. We might find subtle satisfaction in looking at a pattern that 'repeats' in some logical way. Whilst a reflection certainly is classed as a type of symmetry, there are many more types.

In physics, symmetry is usually baked into the construction of an object called a Lagrangian. Lagrangians are mathematical expressions that store pretty much all the information we need about a physical system—they are a system's kinetic energy subtracted by the potential energy of the system, and encode terms to also take into account the interactions between separate bodies in a system. They often contain symmetries, which means that they don't change when we twist and turn the fields that make them up in some particular way.

Symmetries and Lagrangians are special in physics because they allow us to construct conserved quantities. Conserved quantities are physical, observable quantities that stay the same throughout the evolution of a physical system. They are interesting not only because of their philosophical consequences but also because of their use in solving equations. It is easier to solve mathematical equations with quantities that you know sit still and remain constant.

Continuous symmetries are 'smooth' symmetries, like a rotation. Noether's theorem states that for every continuous symmetry, we can construct a conserved quantity. So, for example, if we

*have rotational symmetry in a physical system, we automatically
get conserved angular momentum for free. Surprisingly, Noether's
theorem can show that energy conservation is a consequence of the
time translational symmetry—or when the Lagrangian itself doesn't
depend on time. In other words, if the background scene of which a
physical system is placed remains the same throughout time, then
the combined energy of that system will also remain the same.*

*This concept of symmetry is pervasive in simple mechanics
and every field of classical and modern physics. For example, in
quantum physics, the symmetry of quantum mechanical systems
corresponds to the conservation of quantum angular momentum.
In the theory of electricity, the conservation of charge and spin of
electrons results from symmetries that electrons obey.*

*There are proposed theories which are motivated by symmetries.
However, there are also symmetries which don't hold in nature,
regardless of how appealing they may seem. In the following essay
on charge, parity, and time symmetry, Professor Rigopoulou discusses how the symmetries we once thought were true are no longer deemed to be so. She will then consider one of the most extreme
examples of asymmetry in nature, the arrow of time, and its con-
nection with the idea of entropy and thermodynamics.*

The fundamental second law of thermodynamics states that the
entropy of the Universe always tends towards a maximum. If the
Universe had been born into a high-entropy state, there would
have been no galaxies, no stars, no planets, and no life. Hence,
the primary reason why we are here is the initial low entropy
of the Universe. But entropy is a measure of the 'disorder' of a
physical system. In terms of the underlying quantum description, entropy is a measure of the number of quantum states that
are necessary in order to describe a system in terms of macroscopic variables, such as temperature, volume, and density, all of
these variables increase as the Universe evolves.

Yet, our everyday life is full of 'symmetries'. For most of us,
when we hear the word symmetry, we think about reflecting

images in a mirror. The standard model of particle physics also has three related (near) symmetries, the combination of which is also a symmetry (known as the CPT symmetry).

In this chapter, I will further explore the concepts of entropy and symmetry in the Universe and explain how each of these apparently 'inconsistent' properties has shaped the world we live in.

The Meaning of Symmetry

The word symmetry comes from the Greek word συμμετρια and means the same measure, often referring to items that are 'equally proportioned'. Symmetry is an expression of exact correspondence between things or, in some sense, a measure of indistinguishability. Everyday life is full of such examples and humans experience symmetry from a very young age: babies recognise symmetry in the facial features of their parents and children experience mathematical symmetry when practising additions such as 1 + 2 = 2 + 1. Symmetry also reveals itself in the physical world, the cycle of the seasons and in music in the tones of the songs. In science, symmetry is used to describe the properties of the microscopic and macroscopic world, from atoms and molecules to the structures of the Universe.

In physics, symmetry is an extremely powerful concept. The laws of physics, which govern the observations of what can and cannot happen in the Universe, are a natural consequence of such universal symmetries. Take conservation of energy: in an *isolated* system the total energy remains constant. This principle can never be violated. If this were the case, then the activity in the cells in our bodies would change at any time changing the way our bodies work. The energy released when nuclei fuse would fluctuate, altering the Sun's energy and drastically affecting life on Earth. The symmetry embedded in the laws of nature, such as the conservation of energy, has shaped the nature of our world.

Symmetry also led to the discovery of the cornerstones of matter called quarks [1, 2]. In the early 1950s, the physicist Murray

Gell-Mann looked for regularity in the 'zoo' of particles produced by particle accelerators. In his research, he used the principles of symmetry to predict the existence of fundamental particles, which he named quarks. Physicists first observed quarks in the early 1970s at the Stanford Linear Accelerator Center [3–5].

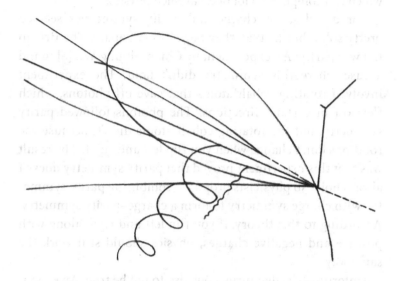

So far, we have talked about the laws of physics and established that they obey symmetries. But is the Universe symmetric? This is a really important question to know the answer to when doing calculations because symmetries tend to make the mathematical description much easier. Most of the physics discovered over the last century, including general relativity and quantum mechanics, is based on three main symmetries: *charge, parity,* and *time.*

Charge symmetry means that we could swap all the Universe's positive charges for negative charges and vice versa without changing anything important. Parity symmetry means we could flip the Universe right to left like a mirror and nothing would change. And time symmetry means that we could run the Universe backwards in time without changing any of the laws of physics.

Of course, not all of these symmetries hold. As we shall discuss in the next section, time symmetry is broken by the laws of thermodynamics, which state that entropy can only increase forward in time. But what about charge and parity symmetries? As it turns out, these symmetries are also broken in specific circumstances, which can complicate a lot of established physics:

For decades, the charge and parity symmetries seemed pretty solid, but in 1956 they began to fall apart. The first to fall was parity: An experiment by Chien-Shiung Wu [6] found a case where this symmetry didn't hold. The experiment involved rotating cobalt atoms that gave off photons, which flew off in certain directions. The photons followed parity symmetry, but the rotating cobalt atoms didn't, because the rotations don't change when you flip left and right. The result was that this experiment proved that parity symmetry doesn't always hold, so physicists proposed combining parity symmetry with charge symmetry to form a charge–parity symmetry. According to this theory, if you flip left and right along with positive and negative charges, physics should still work the same way.

Unfortunately, that turned out also to not be true. An experiment in 1964 involving exotic particles called kaons violated charge–parity symmetry. So physicists had only one option left: to combine charge–parity symmetry with time symmetry to form charge–parity–time symmetry. Surely this symmetry can't be broken, right? So far, it actually looks like this final type of symmetry might be safe. Physicists have been trying for over half a century to break it, and they've been unsuccessful. But all it takes is one instance where charge–parity symmetry doesn't hold, and all of physics might have to be rewritten. Hopefully, this last symmetry stays safe.

The Concept of Entropy

The identification of entropy is attributed to Rudolf Clausius (1822–1888), a German mathematician and physicist. However, it was a young French engineer, Sadi Carnot (1796–1832), who first hit on the idea of thermodynamic efficiency; although, the idea was so foreign to people at the time that it had little impact. Clausius was oblivious to Carnot's work but hit on the same ideas.

Clausius studied the conversion of heat into work. He recognised that heat from a body at a high temperature would flow to one at a lower temperature. This is how coffee cools down the longer it's left out—the heat from the coffee flows into the room. This happens naturally. But if you want to heat cold water to make the coffee, you need to do work—you need a power source to heat the water.

From this idea, Clausius proposed that the entropy of any *isolated* or *closed system* will increase with time [7]. The meaning of the term entropy has its roots in the Greek words 'εν' and 'τροπή', which translate as 'towards conversion', therefore describing the change of energy when moving from one state to another. Clausius' suggestion applies to all *irreversible* processes and is summed up in the second law of thermodynamics: the entropy of an *isolated* system either remains constant or increases with time [8].

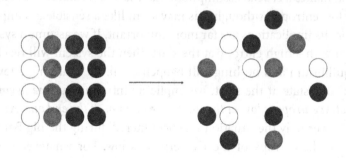

But it was thanks to Boltzmann in 1877 that entropy, a concept whose real meaning is hard to grasp, was linked to the properties of atoms in a macroscopic system [9]. Boltzmann suggested that the exact properties of a macroscopic system can vary considerably; in other words, particular atoms that are *indistinguishable* from our macroscopic perspective can arrange themselves in various ways. Moreover, he suggested that low-entropy objects are more delicate with respect to such rearrangements. The situations that we characterise as 'low entropy' seem to be easily disturbed by rearranging the atoms within them, while 'high-entropy' ones are more robust. So the concept of entropy can now be expressed as a measure of the number of particular microscopic arrangements of atoms that appear indistinguishable from a macroscopic perspective.

This new definition of entropy has far-reaching implications, not least because entropy is no longer a phenomenological concept of thermodynamics but a concept that can be derived from physical principles linked to a macroscopic system. Moreover, it is now clearer why entropy tends to increase in an *isolated system*: because there are more ways that atoms can arrange themselves in a system with high entropy than in one with low entropy.

The Universe: The Arrow of Time and Its Asymmetry

The Boltzmann definition of entropy while simple in its conception makes a crucial assumption: that the system starts in a state of low entropy. Although this may seem like a sensible assumption, its implications are far more important. If we assume a system with a high entropy at the start, then the system will reach equilibrium and nothing will happen at all. By assuming a low entropy state at the start, we implicitly introduce a *time asymmetry: entropy is low at the start state and not at the end one.* And this is exactly the situation we encountered during the Big Bang when the entropy of the Universe was low. For whatever reason, of the many ways we could arrange the constituents of the Universe, at early times they were in a very special, low-entropy configuration.

To deal with this concept of time asymmetry, physicists have introduced the concept of the *arrow of time*. In his book *A Brief History of Time*, Stephen Hawking introduces the thermodynamical arrow of time, as the direction of time in which the entropy of a closed particle system grows, according to the second law of thermodynamics [10]. A classic thermodynamic system consists of a huge number of particles. At the microscopic level, the physical laws governing the motion of an individual particle do not distinguish motion of the future from that of the past. But if we consider the behaviour of the whole system, we will notice that some natural processes never occur in reverse order, even if they do not violate the laws of physics such as the conservation of energy. Take as an example a drop of ink. The drop will diffuse into a glass of liquid; however, we will never observe the reverse effect where the diffuse molecules reassemble spontaneously to form the drop (unless we cause it with some artificial intervention in the system).

The initial drop of ink represents a system of high order (and low entropy), just like the young Universe right after the Big Bang. As time goes on (the arrow of time moves in one direction) the ink molecules diffuse in the water, the order is destroyed, and the entropy of the system increases. So, the arrow of time (the asymmetry) works in such a way that the entropy of the system (or the Universe) increases. It is worth however mentioning that the second law of thermodynamics is a 'statistical law', not an absolute law (like that of the conservation of energy). And although moving from high entropy to low entropy (from chaos to order) is not prevented from a thermodynamical point of view, such a move is very unlikely to happen. If this were the case, then the arrow of time would be reversible but that has never been observed.

The arrow of time is evident everywhere in the Universe: light from the Sun heats the Earth. As a consequence of the second law of thermodynamics, heat flows from the hot object (the Sun) to the colder object (the Earth). If this process was the only thing happening, then there would come a point where the Earth–Sun system would reach equilibrium and the Earth would become a hot and unpleasant planet. Luckily, this does not happen. The reason our planet doesn't reach the temperature of the Sun is that the Earth loses heat by radiating it out into space. This happens because space is much colder than the Earth. Because the Sun is a hot spot in a mostly cold sky, the Earth doesn't just heat up, but rather absorbs the Sun energy, processes it, and radiates any excess back into space. Throughout this process, entropy increases. *All these events* are possible because of the second law of thermodynamics. In other words, entropy and the arrow of time enable life on our planet.

So, where does this leave our original argument about symmetry in the Universe? As Nobel Laureate David Gross remarked, if it weren't for symmetry-breaking the world would be an extremely boring place. In every microscopic examination, you would see the same thing over and over again.

EDITORS' COMMENTARY

Dimitra Rigopoulou has mentioned two key ideas in physics— symmetry and entropy. In terms of symmetry, the hypothesised symmetries of charge, parity, and time are discussed, as well as their failures. On the opposite side, the arrow of time, which relates to the concept of entropy, is discussed as an example of extreme asymmetry in nature.

The symmetries of a physical theory are represented through smooth shapes called Lie groups. In physics, there is a vast landscape of many different types of symmetries present in our world. For example, we have symmetries that give rise to particles, symmetries that dictate the laws of electromagnetism, and symmetries that emerge from complex systems. As mathematicians, it is almost irresistible to try and bring order to this wild-west in attempting to classify all of the different kinds of symmetries we have.

There are some simple ways to start. First, we could begin splitting them out by deciding if they are discrete or continuous. A discrete symmetry is a 'jumpy' transformation, like a reflection in a particular axis. A continuous symmetry is a smooth symmetry, like rotating a circle. A reflection is not a continuous transformation. On the other hand, rotations are classified as continuous transformations.

Smooth symmetries are of scientific interest because they tend to have pleasing properties. One of these properties is that they tend to be associated with geometrical shapes. For example, rotational symmetries can be defined by how many degrees you can rotate an object, and so it is natural to associate this transformation with a shape—the circle. Since we can put a coordinate system on a circle, we call them manifolds.

So, we have symmetry structures in physics that are also manifolds themselves or Lie groups. The concept behind a Lie group is reasonably straightforward. Lie groups are mathematical objects that describe smooth transformations. For example, the symmetry group of rotations of an object is a Lie group because rotation is a 'smooth' transformation. By smooth, it means that I can rotate an object just a tiny bit. But, on the other hand, mutations like reflections don't have this smoothness property associated with them. So you can't reflect something 'just a tiny bit'.

Lie groups are interesting because there are theorems that allow us to classify them. A specific type of Lie group, called a semisimple Lie group, has been classified into several distinct families. The Cartan classification for Lie algebras organises these groups. It turns out we can organise all finite semi-simple Lie algebras over into four infinite families denoted An, Bn, Cn, and Dn with five exceptions.

CHAPTER 3

Darkness, Light, and How Symmetry Might Relate Them

Alan Barr

EDITORS' PREFACE

In Caroline Terquem's chapter, we considered the duality of chaos and synchronicity. Dimitra Rigopoulou then explored the concepts of entropy and symmetry. Now Alan Barr will look at another fascinating set of opposites, light and dark.

We will start by reviewing the theory of special relativity, before considering the physical models of light and dark.

Special relativity postulates two things. The first is that the laws of physics are identical in a particular type of frame, known as an inertial frame. The second postulate is that the speed of light is the same in every inertial frame. To understand what an inertial frame is, let us consider perspectives. The observer's perspective dictates how they experience physical laws. For example, moving around and spinning randomly in a room will cause you to see the physics

DOI: 10.1201/9781003306986-3

of the space differently than if you were standing still. A reference frame is the term given to how a specific observer experiences or notates physical laws, and it is unique to the observer itself.

In a particular reference frame, an observer might have a different system of coordinates to label where objects are as compared to another observer. Consider the case of a train moving at a constant speed. If you are waiting on the side of the tracks, and you watch the train whizz by, from your perspective it will look as if the train is moving and you are standing still. However, from the perspective of a person on the train, it would feel like they are still whilst you are moving. Thus, there is an infinite number of different reference frames from which to view a physical system. However, there is also no guarantee that physical laws are identical in different reference frames. This being so, how is it possible to do physics at all?

An inertial reference frame is the only reasonable frame by which we can do physics. Such a frame is 'stable'. Stability means that any object moving at a constant speed remains at constant velocity unless acted on by an external force. The principle of relativity states that physical laws stay the same in any inertial reference frame. To the best of our observations, the principle of relativity is empirically true.

If someone experiences a set of physical laws in an inertial reference frame on Earth, these physical laws would be the same for someone in an inertial reference frame far away on Neptune, in another galaxy, or at the house next door. Einstein's theory of special relativity states that in an inertial frame, the speed of light is constant. This constant is usually denoted with the letter c, and is approximately 3×10^8 metres per second. The principle of relativity implies that in all reference frames, the speed of light remains the same. Thus, in the example given above, the speed of light is the same whether someone is on the train or standing still on the platform. However, a constant speed is not necessarily true for moving objects that are not light. For example, consider the speed of someone cycling on a bike. The cyclist's measured speed is different for an observer on a train than for an observer standing on a platform.

This is why light appears as a stark manifestation of symmetry: the speed of light is invariant under transformations that take an inertial frame to another inertial frame. In the theory of special relativity, the group of transformations that take inertial frames to other inertial frames is called the Poincaré group. It consists of translations, rotations, and boosts that leave the speed of light invariant. Under this framework, we have bizarre and counterintuitive effects such as time dilation and length contraction.

Here is an example of time dilation in action: Suppose you are standing still, and you wait for an hour before a train whizzes past you near the speed of light, at a speed v. In the time it has taken for an hour to elapse for you, a person on the train has experienced a time of slightly less than an hour. Whilst this effect holds true at all speeds, it only becomes discernible at much higher speeds. If the train were travelling at a third of the speed of light, the person on the train would experience a time of 56 minutes to your hour.

The theory of special relativity and the constancy of the speed of light are not the only examples that link symmetry to light. There is a rich symmetric structure in the quantum mechanical picture of light. We find numerous examples of inherent symmetry in quantum electrodynamics (our current theory of the electromagnetic forces that give rise to photons).

In the following essay, Alan Barr discusses light and its lack of interaction with a species of particle called the neutrino. He then considers the current theories around an entirely different beast— dark matter.

There is much more to darkness than just the absence of light.

The first kind of darkness is the absence of light. Blank out the sun, the moon, and the stars; remove candles, torches, and artificial lights; and the world goes black. No beams of light spread out to illuminate our world. No scattered rays bounce off our surroundings and pierce our eyes. No photons excite the cells in our retinas, tingle the nerves that run to our brain, and paint a picture of the world around us.

But that is not the only kind of darkness. To understand the other kinds, we first need to understand light.

To a physicist, rays of light carry energy and information in physical form. Light rays are interconnected ripples, speeding undulations of electricity and magnetism. They need no medium through which to pass and so travel unimpeded through the vacuum of space. Unbounded in their range, and travelling at the ultimate top speed, they convey to us messages about objects at immense distances and times long past.

The information in light is mostly carried in the spectrum of colours it contains. The different colours in the rainbow are rather like the different musical notes. Red light is the deeper, lower-frequency notes. It comes from longer, slower molecular, or atomic oscillations. Blue and violet are like the high notes and are created from short quick oscillations. Green light lies between, in the middle of the visible spectrum. These colours when combined together make white light, a full electromagnetic cacophony of optical noise.

Beyond the colours of the rainbow, which span a single 'octave' of the optical spectrum, there are at least 80 more octaves stretching to both lower and higher pitches. These are the warm infrared rays that let us feel the heat of the sun, and the longer microwaves that might warm our supper. With the right instruments we can detect light from the lowest notes of radio waves right up to the highest pitched x-rays and gamma rays from nuclear decays.

Even within the single octave of the visible light, there's plenty of variety. The various colours behave differently as they ripple through and interact with the material world. As the sun rises, the first rays glancing at the earth's surface must pass far through the atmosphere to reach our eyes. Dust and gas scatter the blue light away leaving only a red sunrise glow to reach our eyes. Later in the day, with the sun high in the sky, it is that same blue scattered light, seen from a different perspective, that gives the sky its blueness.

Materials too respond differently to different parts of the spectrum. Pigments appear purple or yellow or orange, precisely because they absorb only certain colours and transmit others. The grass looks green because the chlorophyll in its leaves absorbs red and blue light, leaving only the green to be seen. But a black cloak is black because it indiscriminately absorbs all visible light, leaving none to scatter to the eye.

This is the second kind of darkness—the darkness of soot, of the crow's wing, or the blackberry. This darkness is caused by objects interacting with and responding to light. Dark objects are dark because they absorb almost all the light that falls on them. The molecules within them feel the pulsating electric field, absorb its energy, and jump to excited configurations. When they later give out that energy, they do so in parts of the spectrum that we can't see. The light that hits them never gets observed. They look black, almost as if no light had struck them. But it's an imperfect darkness. The reemitted light, though invisible to the eye, is merely shunted beyond the visible spectrum. A crow seen by an infrared camera will still appear to glow brightly. It turns out to be most useful to us that each atom or molecule has its own characteristic set of optical harmonics. We can find out what the particular notes of light are that different materials emit and absorb. Those patterns are unique to each species. Armed with a catalogue of optical harmonics we can work out the properties of distant materials—even those far out in space. The sun and the stars, for example, produce light stamped with the patterns of the simplest two elements, hydrogen and helium. Even though we aren't able to gather a scoop of the superheated plasma from the surface of the sun, we can tell what it is made of by the light it sends us. Understanding its substance in turn helps explain how stars like the sun can burn for so long. The pressure in the centre of the sun squeezes and transmutes hydrogen into helium, releasing vast amounts of energy in the process. That energy is eventually released from its surface as the sunlight we observe.

But stars emit more than just light. A few percent of the energy they produce is instead radiated away as ghostly particles known as neutrinos. These subatomic waifs are created in the same fusion reactions that produce light in the core of the sun. But unlike light, which bounces and scatters around inside the sun, the neutrinos barely interact at all. Unresponsive to either electric or magnetic fields they travel almost unimpeded through the Sun, the Earth, and indeed any observer that attempts to try to catch them. Billions of these neutrinos pass through our bodies every second, unobserved, undetected, and undeflected.

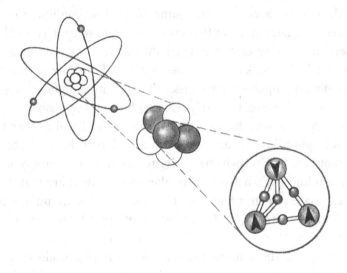

These neutrinos exhibit the third kind of darkness. They are dark not because there is no light to strike them, like the darkness of the night. Nor are they dark because they absorb all the light that falls on them, like soot. Neutrinos are dark because they are transparent. They can't be lit up because light will not bounce off them. This total transparency is not a property of any familiar objects. Any solid, dust, gas, or plasma will absorb some part of the wide range of the electromagnetic spectrum. Glass looks transparent to visible light, but it nevertheless absorbs infrared and ultraviolet light.

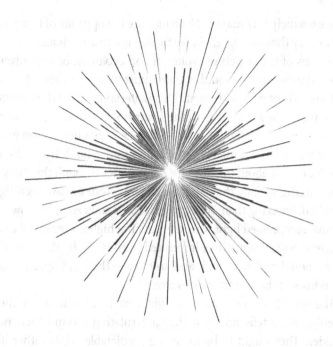

Neutrinos, unlike glass, are transparent to the whole spectrum of electromagnetic radiation. Neutrino means little neutral ones. Being perfectly neutral they just don't feel the oscillating electromagnetic waves of light at all. Light of all colours passes clean through them. Shine a light out into a sea of neutrinos and what you will see is pure unresponsive darkness. This kind of darkness, the darkness of the neutrino, makes them exotically invisible. And a substance that is unseen and non-interacting is hard to study. It provides us with no characteristic spectral harmonics to probe its inner working. No scattered light tells us about its structure and form. We see only darkness and so are both literally and metaphorically unenlightened about its nature.

With the neutrinos we have managed, eventually, to find other ways to understand them. These are techniques that don't rely on interactions with light. Over the past few decades, we have made precise measurements of their feeble interactions and started to tease out some of their secrets. One of the things we have found is that they have extremely tiny masses. Neutrinos are the opposite of heavy, but let's not say they are 'light', lest we further confuse the issue.

Studies of the structure evolution and evolution of the universe have led us to understand that there must be some other substance that doesn't interact with light. This material is called *dark matter*. It is dark in the same sense as the neutrino, in that it simply seems not to interact with light. It is mysterious for the same reason too— things which do not interact are hard to understand. Ironically we only found out about the need for dark matter through the study of light—starlight. Stars that are moving away from us have their light stretched, lowering their optical pitch. Those stars zooming towards us have compressed light waves, producing higher notes. The characteristic optical signatures imprinted in the light by the stars also get squeezed and stretched, and so encode the star's speed in the harmonics of the spectra we observe.

The stretched and squeezed light that reaches us from these speeding stars tells how fast they are rotating around their host galaxies. They ought to be following predictable orbits rather like those of the planets orbiting the sun. The speed of their trajectories depends on the amount of mass that's drawing them in towards the centre of their galaxy. But the inferred speeds are far too high to be explained by the gravitational pull of the other stars alone. Even when stray dust and gas are added to the equation, the peripheral stars are travelling at speeds that should shoot them out into space.

The simplest explanation is that some extra mass is drawing them in. The additional matter does not emit light, and so became known as *dark matter*. It's not dark in the sense of the absence of light, since it is surrounded by luminous stars. Nor is it the darkness of soot, since it neither absorbs nor emits light in any part of the spectrum. Its darkness is non-interacting transparent darkness, like that of the neutrinos. But unlike the neutrinos, dark matter has enough mass and gravitational tug to hold galaxies together.

Other observations have shown that dark matter also pulls on large clusters of galaxies, directing their formation and evolution. Dark matter was present before the galaxies were born. And right in the early universe, before the first stars, before even the first atoms condensed out of the hot plasma of the big bang, dark matter was there, tugging on the proto-atoms. All these observational clues point in the same direction. Dark matter is a heavy invisible material, quite unlike anything we have detected so far on Earth.

This material is dark in the physical sense that it has no interactions with light, but also dark in a metaphorical sense. We know very little about what it's made of. Indeed, the only thing that we really know (other than its transparency which makes it dark) is roughly how much of it there is. The total amount, when added up across the universe, is rather a lot. In fact there seems to be several times more mass of dark matter in the universe than all of the matter contained in all the stars, dust, and gas combined.

Our state of almost complete ignorance about dark matter doesn't stop imaginative and creative people from dreaming about what it might be made of. It's a fruitful field of study and speculation. It turns out that there are very many different ways in which one can explain theoretically something about which there is so little empirical knowledge.

Amongst these many theories of dark matter, some of the most striking relate it by symmetry to other particles that we already understand very well. Indeed in some of the most interesting theories, dark matter is not so different from light itself. These latter theories involve various types of mathematical or geometrical symmetry that directly relate dark matter to light. It might seem odd that dark matter, which is by definition dark, might be closely related to light. But, counter-intuitively, light does share some of the properties expected of dark matter. Rays of light pass straight through one another, without interacting, colliding, or reflecting from one another. The little packets of energy that make up light, known as photons, shoot straight through one another. Light is transparent to light. And in this sense, it shares the property of transparent darkness that we previously encountered in the neutrino.

One class of these speculative theories suggests that dark matter is a form of excited photons with their energy stretched out into other invisible dimensions of space. This excited light would gain mass, travel slowly, and look very much like dark matter. A second class of theory says that dark matter is made of a partner of light called the photino. In this scenario the dark cousin of light is made heavy through the breaking of a symmetry that relates matter to force. The photino would also share light's natural property of transparency. Unlike light it would be heavy and sluggish due to an imperfection in the symmetry. This second type of theory is called supersymmetry, and it too could be the perfect explanation for dark matter.

For the moment, these theories are only speculation. Dark matter and light might indeed be related by a deep symmetry—but we won't know until we can study dark matter and its relationship with light more closely.

Efforts are currently under way at the Large Hadron Collider near Geneva in Switzerland to produce and record the presence of man-made dark matter. Other experiments are seeking it in mines deep underground. The scientists and engineers working on these are attempting to detect the rare collisions between atoms and the relic particles of dark matter left over from the big bang. Other experiments again are looking out into space, searching for the light that might be emitted on the very rare occasions that pairs of dark matter particles collide, are annihilated, and yield up their energy in visible form.

None of these terrestrial experiments has yet been able to observe and identify the dark matter that we know—from the gravitational pull of its mass—must exist. But on they search, in hope and expectation.

Perhaps soon we will start to understand the nature of dark matter. With patience, we might get to know it well—even as well as we do the once-mysterious pinpricks of light in the night sky that are the stars. Our minds, not constrained to understanding only those things that can be seen, could then gain insight into both the light and the dark matter in the universe.

EDITORS' COMMENTARY

Alan Barr's essay is a fascinating exposition about one of the clearest examples of duality in nature, namely light and darkness. Current models in physics frame light itself as a manifestation of symmetry—in special relativity, the speed of light is an invariant quantity under Poincaré transformations. However, special relativity is not the only example. Quantum electrodynamics combines special relativity and quantum mechanics, and it is currently the best model we have to describe the interactions between light and charged matter. The Lagrangian in quantum electrodynamics has a symmetry which is described by a group dubbed U(1)—the group of rotations in a unit circle.

The connections between light and dark matter that Professor Barr discusses are even more surprising. The hypothesis that dark matter is related to light through a more profound symmetry is both counterintuitive and aesthetic. It is important to note that whilst this duality would certainly be beautiful, without experiments we do not know that it holds true.

The immediate association with the word 'chaos' is a negative one, but entropy can act as a highly effective selection principle in the initial conditions of biological systems. For example, it has been demonstrated that biological systems use a chance to increase their likelihood of survival. Denis Noble and Anant Parekh will explore this idea further in their respective chapters.

CHAPTER 4

Self-Similar Self-Similarity

Joel David Hamkins

EDITORS' PREFACE

Thus far, the essays in this book have explored the role of symmetry and order in physics. Let us now turn our attention to the field of mathematics. Whilst the lines between mathematics and physics are somewhat blurred, the art of classifying symmetries into rigorous and well-defined objects has traditionally been the domain of mathematicians. The study of the structures that represent these symmetries is known as group theory or abstract algebra. Group theory is a wide-reaching field of rich research with applications in fields as diverse as physics, computer science, geometry, number theory, and topology.

Groups are mathematical objects that represent transformations. The most intuitive transformations are those that act on shapes, such as the rotation of a square. However, these transformations also apply to abstract spaces or mathematical sets. In terms of basic questions about groups, there are a number of

DOI: 10.1201/9781003306986-4

interesting questions we might consider about the nature of groups. One of these questions concerns size.

For example, how big is the set of symmetries of a triangle? It is not difficult to convince oneself that the set of transformations of an equilateral triangle consists of 0-, 60-, and 120-degree rotations, as well as three reflections. Hence, the group of triangle transformations has a size of six. Or one might ask if there are any 'subsets' of groups that form a group in their own right.

There are several types of groups that are special since they index the transformations of objects we are all familiar with. So, there are specific names we have for them. The set of symmetries that leave an n-sided polygon unchanged is called the dihedral n-group. This group is the set of rotations, as well as the reflections for this polygon. For example, the transformations of a square are part of the D_4 group, which consists of four reflections across the vertical, horizontal, and diagonal axes, as well as the rotations of a square. This is something that will be elaborated upon in the following essay. The group of rotations on its own is dubbed the cyclic group. There is also the symmetric group—the symmetric group of order n is the group of one-to-one functions between a set of n objects.

There are indirect ways to measure the sizes of a group, too. These ways often involve interesting dynamics concerning how a group acts on an object itself. We even have mathematical terms which describe the different states that parts of an object remain the same under transformation. Thus, if I reflect a square through a vertical line down the diagonal, then the two corners lying on that axis remain unchanged.

The orbit of an element is the set of all places a group can take a particular point in an object. For example, the orbit of the top-left corner of a square is every other corner. The orbit stabiliser theorem tells us how we can relate the sizes of a group to the orbit and stabiliser.

In the following essay, Joel David Hamkins discusses groups in general before going on to examine one of the ultimate symmetries, self-similarity. Self-similarity is invariance under a change of scale. For example, does zooming out from a shape look like nothing happened at all? This chapter is a deep dive into some of the more mathematical aspects of symmetry. Unlike the previous chapters, we will be less concerned with the relationship between symmetry and asymmetry or with the notion of opposites, but we will focus on symmetry as a standalone concept.

Let me tell a mathematician's tale about symmetry. We begin with playful curiosity about a concrete elementary case—the symmetries of the letters of the alphabet, for instance. Seeking the essence of symmetry, however, we are pushed toward abstraction, to other shapes and higher dimensions. Beyond the geometric figures, we consider the symmetries of an arbitrary mathematical structure—why not the symmetries of the symmetries? And then, of course, we shall have the symmetries of the symmetries of the symmetries, and so on, iterating transfinitely. Amazingly, this process culminates in a sublime self-similar group of symmetries that is its own symmetry group, a self-similar self-similarity.

In the light of symmetry consider a capital letter A.

Well, this particular A in this particular font, unfortunately, is not quite perfectly symmetric. The uprights at left and right differ in thickness, for example, and the serif on the right foot is ever so slightly larger than the serif on the left. Let us try to draw a somewhat more symmetric letter A, even though it may be less graceful.

That's better. This A exhibits a vertical-line symmetry—the vertical line down the centre.

If we reflect the letter across that line—like Alice through the looking glass—it lands perfectly upon itself. We might fold the paper on that line to realise the symmetry. The letter B also exhibits symmetry.

B

Well, again, this particular B in this particular font is not *perfectly* symmetric—the upper curved half is slightly smaller, and the two curves are not exactly the same shape. But we can try to draw a more symmetric B, if less graceful, so as to exhibit a horizontal line symmetry.

We might similarly consider all the letters of the alphabet, drawing each of them as symmetrically as possible. Get some paper and try!

A B C D E

What symmetries did you find? Some letters, as we have seen, have vertical or horizontal line symmetries. The letter H has both vertical and horizontal line symmetries. The letter S has no line symmetries, but it has a *rotational* symmetry.

If you rotate this letter S by 180 degrees about its centre, half-way around, it will land precisely upon itself. In elegant fonts, the letter S is often not quite symmetric in this way—the upper part is often gracefully smaller than the lower part. Some letters, such as F, G, and R, seem to have no nontrivial symmetries at all.

Which is the most symmetric letter? The letter X, when drawn with perpendicular lines, exhibits not only vertical and horizontal line symmetries but also diagonal line symmetries, as well as fourfold rotational symmetry.

One particular letter, I claim, can be drawn so as to exhibit *infinitely* many symmetries! Consider the letter O, drawn as a perfect circle.

This letter O has line symmetries with respect not just to vertical and horizontal lines, but with respect to any line through the centre at all, infinitely many. And it also has infinitely many rotational symmetries, by any angle you like.

To illustrate how the symmetries of a figure form a mathematical system, let us consider the symmetries of a square.

The square exhibits a four-fold rotational symmetry. If we rotate by a quarter turn, the square lands upon itself. We can rotate twice, or three times, but rotating four times brings back the original orientation. This is the same result as with no rotation, the *identity* symmetry, a trivial symmetry leaving the object unchanged. Perhaps the trivial symmetry is hardly a symmetry at all. Yet mathematicians systematically find it useful to regard trivial or degenerate instances of their conceptions as fully valid—every square is also a rectangle; every equilateral triangle is also isosceles; and zero is a number. So let us regard the identity symmetry as a symmetry—it is one of the ways of associating the square rigidly with itself.

The square also exhibits symmetries by reflection, with four lines of symmetry.

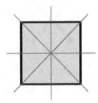

Thus, the square has eight symmetries in all: four rotational symmetries (including the identity symmetry as 0-degree rotation) and four reflection symmetries.

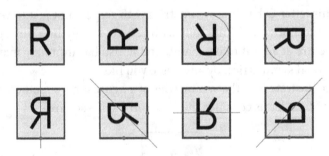

The letter R indicates each symmetry and how it acts upon the square.

Symmetries can be *composed* by performing them one after the other; this is a kind of multiplication operation for symmetries, with the result being another symmetry. Thus, symmetries become dynamic—each is a distinct transformation. If you first reflect on the vertical line and then on the horizontal line, for example, which symmetry have you got?

Altogether, this is a rotation by 180 degrees, halfway around. The composition of two reflections in the plane is always a rotation, because the mirror orientation, being twice reversed, is ultimately preserved.

Every symmetry admits an *inverse* symmetry, the symmetry that undoes it; composing them is the identity symmetry. The inverse of clockwise rotation is a counterclockwise rotation through the same angle; reflections are self-inverse since performing them twice gives the same result as doing nothing.

When composing symmetries, does the order matter? Yes, indeed it does. If you compose a left-quarter turn with a vertical-line reflection in both ways, the end result is not the same.

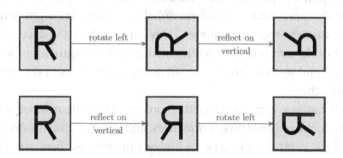

Abstracting to a higher dimension, consider the symmetries of a cube.

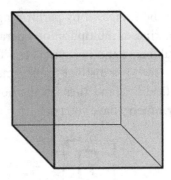

There are numerous rotational symmetries—grab the cube by any face and apply a quarter turn, or grab the cube by a vertex and spin by one-third, or grab the cube by an edge and twist exactly halfway around. There are also numerous mirror-plane reflection symmetries—how many can you find? And there is the *central symmetry*, turning the cube inside out by exchanging each vertex with its opposite through the centre. This is neither a rotation nor a planar reflection, it turns out, although it can be realised by composing a rotation with a planar reflection.

How many symmetries of the cube are there? You might be surprised to learn that there are 48 distinct symmetries of the cube. To count them, consider how a symmetry might act upon a particular face. This face must be carried to one of the six faces, and as we observed above, there are eight ways to associate the two squares. Having attached the face, the rest of the symmetry is determined (perhaps requiring reflection), and so there are 6 × 8 = 48 many symmetries of the cube altogether.

What is a symmetry? With geometric plane figures, we had rotational and reflective symmetries, and in three dimensions we had the central symmetry. In higher dimensions, there are still other kinds of rigid transformations. All of these symmetries are *isometric*, which means that they preserve distances—they do not stretch or compress the figure.

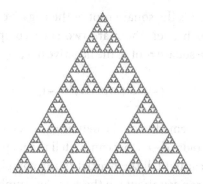

But some figures are insightfully described by nonisometric symmetries. Consider the Sierpinski gasket fractal here, for example. Notice how it appears within itself in scaled-down form—the whole triangular figure appears three times at half size (half the edge size), once in each corner. And it appears nine times at one-quarter size, 27 times at one-eighth size, and so on. There are infinitely many scaled-down copies of the whole fractal inside itself. This is a kind of symmetry, to be sure, a self-similarity, but not by rotation or reflection; because of the scaling, it does not preserve distances. Yet the figure is insightfully described by this self-similarity.

So we are pushed toward a more generous, abstract conception of symmetry. We might consider any structure-preserving transformation of a mathematical object, an isomorphism of the object with itself, as a symmetry of that object. We may consider the symmetries of any mathematical structure at all.

Consider the system of complex numbers C, for example. Complex numbers have the form a + bi, where a and b are real numbers and i is the *imaginary unit*, the square root of negative one,

$$i = \sqrt{-1}$$

We can add complex numbers and multiply them, and altogether, the complex numbers form a mathematical structure known as a *field*.

We said that i is the square root of the negative one. But suppose I ask, which one? There are two such complex numbers, since $-i$ is also a square root of the negative one, as you can check:

$$(-i)^2 = (-1)^2 i^2 = i^2 = -1.$$

So what had seemed to be the defining property of i is a property that also holds of $-i$. How can we tell them apart?

Indeed, we cannot tell them apart in the complex field. There is nothing you can say about i in the complex number field C that isn't also true of $-i$. Your i might be my $-i$, for all we know, if we treat the complex numbers strictly as a field. The reason is that there is a symmetry of the complex numbers, an isomorphism of the complex field with itself, that swaps the numbers i and $-i$. This symmetry is called *complex conjugation*, and it associates every complex number with its complex conjugate:

$$a + bi \qquad \longmapsto \qquad a - bi.$$

The numbers i and $-i$ therefore play exactly the same structural role in the complex number field. They are perfectly symmetric copies of each other.

There are many other automorphisms of the complex field—an enormous uncountable infinity of them, although one uses the axiom of choice to prove this. Every irrational complex number can be moved. And yet, all these symmetries of C, including complex conjugation, are broken by augmenting the complex field with its coordinate structure, using the real and imaginary parts. Thus, the complex plane (as opposed to the mere field) is *rigid*—it has no nontrivial symmetries.

Let us continue our flight toward abstraction. Start with any mathematical structure at all—a geometric figure, a number system, whatever you like—and consider its symmetries. Collectively these constitute the *symmetry group* of your structure. Since this symmetry group is a perfectly good mathematical structure of its own, with composition as the group operation, we may consider its symmetries, or in other words, the symmetries of the symmetries of the original structure.

But why stop there? This next symmetry group, after all, also stands on its own as a mathematical structure, with *its* symmetry group, the symmetries of the symmetries of the symmetries, and then of course there will be the symmetries of the symmetries of the symmetries of the symmetries, and so on. We may iterate the process as long as we like. In this way, we are led to the *automorphism tower*.

$$G_0 \to G_1 \to G_2 \to G_3 \to \cdots$$

We began with the symmetry group G0 of the original structure, and then each next group is the symmetry group or automorphism group of its predecessor.

It is an elementary fact in group theory that every element of a group generates an inner automorphism of that group by the process of conjugation. Because every group element is thus mapped canonically into its symmetry group, the tower of groups can be seen as building toward a certain limit group G_ω, the direct limit of the system of groups.

$$G_0 \to G_1 \to G_2 \to \cdots \to G_\omega$$

And not only that. Because G_ω is a perfectly good group on its own, we may consider its automorphism group, and the automorphism group of that group, and so on, thereby continuing the automorphism tower beyond infinity.

$$G_0 \to G_1 \to G_2 \to \cdots \to G_\omega \to G_{\omega+1} \to G_{\omega+2} \to \cdots \to G_\alpha \to \cdots$$

Iterating transfinitely, each next group is the automorphism group of its predecessor, and at limit stages we use the direct limit.

Does it ever stop? Will there ever be a group that is isomorphic to its own automorphism group by the inner-automorphism association we described?

Amazingly, the answer is yes. The process eventually reaches completion. In my article [11], building on key earlier work of [12], I proved that every group has a terminating transfinite automorphism tower.

What I proved is that in every automorphism tower, perhaps very far out in the transfinite part of the tower, there will eventually be a *complete* group, a group for which every automorphism is already realised as an inner automorphism by a distinct group element. Such a group thus exhibits a perfectly self-similar self-similarity. Nothing new is added beyond this group by considering symmetries of symmetries, or symmetries of symmetries of symmetries; one already has them all.

Since every automorphism tower is completed in this way, if you iteratively consider the symmetries of a structure, and then the symmetries of the symmetries, the symmetries of the symmetries of the symmetries, and so on, iterating transfinitely in the natural manner, then you will eventually achieve a complete, sublime group of self-similar self-similarity. You will eventually, perhaps transfinitely, complete the process of symmetry.

EDITORS' COMMENTARY

One intriguing aspect of this essay is its examination of the symmetries of symmetries. In mathematics, these sorts of meta-transformations are often just as interesting as the original symmetries themselves. Take an object, and then tally up all the ways that it can be transformed. The transformations that leave this object unchanged form a mathematical group themselves. This group is called the automorphism group—it is the set of symmetries that apply onto an object. An automorphism group is a group in its own right. Since it is a group in its own right, we can look at automorphisms of the automorphism group, and so on.

Another interesting angle concerns maps between different groups that preserve some structure. For example, consider the set of transformations of a square as one group on its own, and the set of shuffles of the set {1, 2, 3, 4}. What kind of relationship do these two groups have with each other? Is there some relationship that preserves the group structure between the two? This is called a group homomorphism, and it allows us to determine when groups arising from different contexts are actually 'the same'. We will defer to the technical details about what group homomorphisms are to the reader.

Automorphisms occur when there is a symmetry between symmetries, and this bears some relationship to the symmetries we find in fractal-like patterns. The symmetry of a fractal is somewhat distinct, yet it still holds a very important place in theoretical physics. This is a symmetry of scaling, and is similar to the idea of the renormalisation group in physics. Renormalisation is the study of how the parameters of physical theories change when copies of physical theories are examined under different energy scales.

Professor Hamkins' extraordinary essay reminds us of a metaphor from Hindu mythology. Indra's Net is a vast web of silken strands that stretch out to infinity in every direction. Where the threads of the net cross over one another, at the intersection of the warp and the weft, there hangs a crystal-clear diamond. If you look closely at any one of these diamonds you will see symmetries reflected within it, and symmetries of the symmetries, and so on to infinity—a sublime self-structure of symmetries!

CHAPTER 5

The Language of Symmetry in Music

Robert Quinney

EDITORS' PREFACE

The previous essay focused on groups as mathematical objects which represent symmetries, and we find the same mathematical structure in music. For example, there is rotational symmetry in the circle of fifths, much like how the rotations of a circle, or how the integers modulo a prime, form a cyclic group. This rotational symmetry is only the tip of the iceberg: in our discussions on group theory and shape transformation, reflections have also been used to generate variations from themes as early as 14th-century pieces, an example of which will be highlighted later. Professor Quinney applies these concepts of symmetry to ideas in harmony, structure, and tonality.

Notes pass quickly away; numbers, however, though stained by the corporeal touch of pitches and motions, remain.

[13]

DOI: 10.1201/9781003306986-5

How might we define musical symmetry? We could begin with some examples of order in music—the internal organisation of pieces of music, their 'form' or structure—and see whether or not these match the definition of symmetry: *transformations that leave the object unchanged*. We could fetch some scores from the library and look for evidence.

Music, however, cannot be reduced to a text. It is also, in fact it is primarily, an act [14].

So simply *looking* at music will not help us: we need to proceed with ears open as well as eyes, asking what the music's effect on us might be as listeners.

One more caveat: the examples here are all situated within the Western art music tradition, what is popularly known as 'classical music'. The music under discussion here has all come down to us via documents, the existence and survival of which depended upon their association with high social status and wealth. While it is probable that symmetry can exist in all music, here I focus on the music of a specific tradition—one which, due to globalisation, continues to be a pervasive influence in millions of lives [15].

Symmetry is most often thought of by mathematical dunces like me in terms of reflection. This is only rarely how symmetry manifests in music, however. Why? Most music contains some sort of *narrative*: not songs only, but 'abstract' music too. Music unfolds in time, but not as a succession of unrelated moments. It isn't simply one thing after another; rather in music, one thing usually *leads* to another. Lift the lid of most musical works, and what is revealed is an ordered collection of mutually reliant constituent parts, whose relationship to one another is defined by their *successive* nature (or, sometimes, a rhetorically unsuccessive gesture like a sudden pause or a surprising chord). The principal medium for this successiveness is *harmony*, the means by which melodies are simultaneously combined with one another—everything from a solo voice over a bass line to the densest orchestral score. And because it is virtually impossible to narrate backwards, the uses of mirror (reflection) symmetry in music are limited. We shall look further at narrative in music, and the ways in which symmetry helps to generate it. But not before we encounter two pieces that *do* employ mirror symmetry, both for the delight of an audience of *cognoscenti* and to demonstrate the composers' ingenuity.

The clue is often in the title: never more so than in this *rondeau* by Guillaume de Machaut (d. 1377), *Ma fin est mon commencement*. Those performing it in the days before modern editions (see Figure 5.1) would have needed to take the text to heart: in particular the line 'Mes tiers chans trois fois seulement se retrogade et einsi fin' ('my third voice reverses itself three times only and thus ends'). For a start, there are only two notated voices, and one voice has only half the notes of the other. The solution is for one singer to read from the shorter part, twice: once forwards, then backwards. This part is a reflection of itself, one 'side' of the reflection sung after the other. Meanwhile, the second and

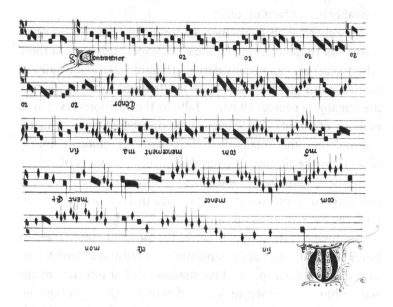

FIGURE 5.1   Machaut, *Ma fin est mon commencement*. Paris, Bibliothèque Nationale, MS Fonds Français 9221, f.136.

third singers share the written-out music, but they read at 180° to each other, one ending where the other began. One reads right-way-up, the other upside down: in other words, these two parts are a simultaneous reflection of each other, designed to sound simultaneously.

Nearly 200 years after Machaut's death, a joint publication by Thomas Tallis and William Byrd included as its final number another tour-de-force of musical symmetry. It takes a short text, 'Miserere nostri, Domine' (Have mercy on us, Lord), and spins a seven-voice web: two canons arrayed across six voices, and one 'free' part. The upper two voices, called *Superius* and *Superius secundus*, have a straightforward canon of the *Frère Jacques* sort: they sing the same melody at a temporal distance. We can hear the relationship clearly, since the two voices are not far apart temporally, and sing at the same pitch: had theirs been a canon at the fifth (with the second voice singing the melody five notes higher), our brains would probably have given up on it.

The other canon is another matter. Below the two Superius voices sits Discantus, singing the *dux* (leader) from which three other voices are canonically derived (Figure 5.2). Incidentally, in the *Discantus* partbook this piece is attributed to Byrd, whereas the remaining sources all name Tallis as the composer: evidence, perhaps, of a joint effort whereby one contributed the basic material and the other 'realised' it. This four-part canon is doubly obscure to the listener. It is arrayed across two pairs of voices: *Discantus–Contra Tenor*, and *Bassus–Bassus secundus*. The lower pair sings an inversion of the *Discantus* melody: where the *dux* falls, the melody in the *Bassus* pair rises by the same degree: a mirror image of the original. Unlike the *Superius* canon, the four lower voices all begin at the same time, but move at different, proportionally related speeds. Like shadows cast at intervals by the setting sun, their dimensions are different, but all are proportionally related to their source. *Contra Tenor* sings the *dux* in double augmentation—each note is four times the length of the original.

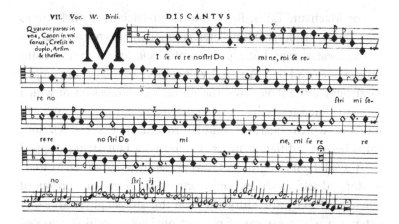

FIGURE 5.2   Tallis/Byrd, *Miserere nostri* (*Cantiones Sacrae*, 1575), Discantus part. Note (a) the attribution to Byrd, unique to this partbook, and (b) the symbols [X] showing where each of the three parts that sings a rhythmically augmented transformation of these notes comes to an end. All three parts derived from this one sang from their own partbook, with a realisation of this canonic *dux* provided (and attributed to Tallis).

Meanwhile, *Bassus secundus* sings the *dux* at half the original speed, and *Bassus* [*primus*] in triple augmentation, at eight times the original duration. All this is well beyond our cognitive ability: indeed, can we experience this canon at all, or is it just a conceit to amuse the knowing reader?

In the *Superius* canon we can hear the music travelling forward. The fact that the second voice literally follows the first, their phrases overlapping, lends a gently propulsive quality to the arrangement. By contrast, the complex duration relationships of the *Discantus* canon (which we could represent as 8:2:1:4, working from the highest to the lowest voice) give their music a quality we might characterise not as two parallel straight lines, but as four concentric circles. This sense of circularity is a common characteristic of complex canons. Time seems to be standing still—or, at least, moving rather more slowly than usual. This music is not, as it were, normal. Why?

For Machaut, Tallis, Byrd, and for musicians of their times and some considerable time later, composing was not an exercise of the creative ego, but a process of *inventing*—from the Latin *invenire*, to 'find out'. Any individual piece of music was subject to ancient immutable laws, the numerical expression of which was attributed to Pythagoras. *Pythagorean* has come to refer to ratios by which musical sound is ordered: that is, the relationships between different notes. Famously, a taut string stopped halfway along its length, and thus vibrating at a ratio of 2:1 to its unstopped self, produces a note one *octave* above that produced when it is full length. Outward from this simple physical reality extends a whole system of proportional relationships, adopted by theorists from Boethius (d. c.524) onward to explain musical phenomena of all kinds. Music, to this worldview, is a fundamental constituent of the created universe: it is sounding evidence of the perfection of that creation. Music was literally everywhere. Following Plato, Boethius proposed three types of music: of the universe, of the body, and finally the music actually produced by human activity.

In order to make their intricate canonic constructions work according to the conventions of their time and place, the composers we have so far considered had to favour *consonance* and mostly eliminate *dissonance*. The relationships between two sounding pitches—their relative frequencies, what we call the *intervals* between notes—were codified according to ideas of perfection and imperfection. We are dealing here with *harmony*. Its roots are in the Pythagorean numbers, and specifically the *harmonic series*: the pitches that resonate, in unvarying order, above the 'fundamental' pitch of any note struck, plucked, blown, or sung. The two are interconnected: the first harmonic is the octave (2:1), and the interval of an octave (or two octaves, or three, and so on) is considered *perfect*; likewise the interval of a fifth, whose ratio to the fundamental is 3:2, and which appears third in the harmonic series (in fact as a 'twelfth' i.e. an octave plus a fifth above the fundamental). Less perfect but still *consonant* were the third and sixth, and these two turn out to be helpful in the composition of music that has more than two parts: imagine two notes a third apart, and move the upper note down an octave, it becomes a sixth, and *vice versa*. In other words, these intervals are *invertible*. All other intervals were *dissonant*: they were literally unharmonious. But without them, music was just one consonance after another, unvarying, and bland.

How was dissonance released into the musical stream without polluting it? There were two methods. First, dissonances could *pass*, unnoticed, between adjacent consonances. Second, and more importantly for our purposes, dissonances could be hit head-on, *accented*, before *resolving* into consonance. A 'passing note' is—no surprise—a form of passing dissonance, while an *appoggiatura* and a 'suspension' fall into the category of accented dissonance.

Ma fin est mon commencement and *Miserere nostri* work by avoiding accented dissonance entirely. There are passing dissonances, gently brushing the consonant surface, but nothing that must be *resolved*. Accented dissonance is prominent: we notice that it doesn't fit. It poses a problem that requires resolution, and convention dictates that the resolution happens in a fixed temporal and melodic direction: it must resolve forward in time and downward in pitch. To do this in reverse would create chaos: dissonance springing from nowhere and left hanging (like the end of the sentence spoken as if it were a question?). Just as only certain dispositions of letters will create a palindrome, only a limited number of rhythms and pitches can be deployed in a complex canonic array. For the most part, it might seem that symmetry was ruled out of this conception of music.

It is certainly true that symmetry came into its own when the system described above, which had endured since before the days of Tallis and Byrd, collapsed around the turn of the 20th century. As the dust settled, composers looked for an alternative set of

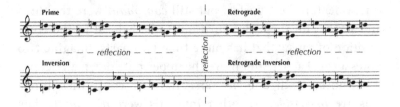

FIGURE 5.3   An example of a 'tone row' and its basic permutations. The sounding pitch may be changed (e.g. all notes of the Prime form moved up three semitones, the shorthand for which is P3), but the *intervals* between the notes must remain the same (e.g. in P3 the first note would be F natural, the second E, the third B). This 'row' and its permutations provide all the melodic material for a piece of music.

conventions—a new framework for their music—and symmetrical transformations acquired a new and prominent role.

In the early 1920s, Arnold Schoenberg sought a language for music that would replace the old system with something equally rigorous and 'universal'. The 12 notes of the 'chromatic scale' would now have equal status, and order imposed by using an unvarying 'row' or 'series' of all 12 notes, in any order, without repetition [16]. Crucially, the row would be subject to transformations, including reflection (either inversion or retrograde motion, or both) and translation (in musical terms, transposition so the row begins on a note different to the 'prime' form: see Figure 5.3). Here was a new musical language, spoken most eloquently, at first, by Schoenberg and his pupils Anton Webern and Alban Berg: the triumvirate often referred to as the 'Second Viennese School'. It says something about human creativity that these composers, all using the same system, each produced such astonishingly individual music.

Of the three, Webern was arguably the most concerned with the internal coherence of his music. In a revealing doodle, he inscribed one of his compositional sketches with this square palindrome:

S	A	T	O	R
A	R	E	P	O
T	E	N	E	T
O	P	E	R	A
R	O	T	A	S

The 'sator square' could be a composition by Webern, if we replaced the letters with notes (except he had 12 notes at his disposal, not 8 letters). His music speaks of a belief, not uncommon in the mid-20th century, in order as an end in itself; specifically, of music purged of decadent appeals to emotion. After his death in 1945, his influence spread across Europe and the United States, and an early standard bearer was Pierre Boulez, who wrote an article entitled 'Schoenberg est mort' (indeed he was, but this was no admiring obituary). Boulez's teacher Olivier Messiaen had introduced serialism to the manipulation of rhythm as well as pitch. Now Boulez included dynamics (relative loudness), and 'attacks' (e.g. how sharply a piano key was struck and released) in an ultra-constructivist system known as 'total serialism'. Series of pitches, rhythms, dynamics, and attacks were allotted numbers and fed into matrices that more-or-less instructed the composer how the piece should turn out. In other words, this was a composition by algorithm—a conscious repudiation of 'romantic' ideas of artistic autonomy. Boulez later reminisced that he had wanted 'to strip music of its accumulated dirt and give it the structure it had lacked since the Renaissance' [17].

Had music come full circle? Hardly, for Boulez and his colleagues' interest in the music of 'the Renaissance' resided in the ways in which it was controlled by 'structure'. In reality, the conventions of harmony, of consonance and dissonance, were far broader and more permissive than the straightjacket of total serialism. The two strictly canonic pieces we have considered were quite unlike the vast majority of music produced by Machaut, Byrd, and Tallis. Symmetry is there even in less outwardly symmetrical music—indeed, we might even hold it responsible for some of the 'dirt' decried by Boulez. And so to Johann Sebastian Bach (whose initials were arranged in near-symmetry for his seal—see Figure 5.4).

But first, more Pythagoras. The system of *tonality* employed by Bach derived from the same numerical relationships explored above. Each tonal area—what we would call a *key*—had as its boundary an octave (2:1). We call the first note (or *degree*) of the scale the *tonic*. The fifth degree of the scale—3:2 to the tonic—is called the *dominant*. In 'tonal' music, the tonic and dominant exist in creative tension. Tonal music begins in the tonic key, and ends in it too, with the tonic placed reassuringly at the bottom of the final chord.

This system does not give the appearance of symmetry. If the octave (between, say, a note C and the C above) were divided exactly in half, the mid-point would not be a fifth above the lower note. Imagine moving inwards from an octave played on a piano keyboard. The mid-point is F sharp, or G flat. In Pythagorean terms, however, those notes are not one and the same: it is only by dividing the octave into 12 equal semitones that such an 'enharmonic' equivalence is made possible, and if we divide things that way, the

FIGURE 5.4 Centre: J. S. Bach's seal, designed in 1722. To the left and right: the initials J, S, B extracted from the seal, demonstrating their near-perfect reflection symmetry.

ratio between the tonic and dominant (C and G) is not a perfect 3:2. There is no enough space here to plot the history of tuning systems, but we need to unlock this apparent problem that the dominant is not produced by a symmetrical, 50/50 division of the scale. The key to this lock is the idea of *inversion*—or, more properly, *invertibility*.

In the first of his two-part *inventions*, Bach shows us how we might 'find out' a modest piece of music (Figure 5.5). First, an assortment of notes is played by the right hand: the *subject*. It is immediately answered by the left hand, which plays the same subject an octave lower; meanwhile, the right hand continues with a complementary set of notes above. Note that the subject begins on C, the tonic, and rises to G, the dominant: by the time the left hand reaches this latter point at the start of bar 2, the right hand has risen to D, which is a fifth above G and therefore *it is* dominant. The listener senses that something has changed: we're no longer where we started. Sure enough, we now hear the subject again, in the same exchange between the hands, but starting this time on G. Very soon after beginning, the music has reached the opposite pole of the tonal planet, the dominant. All that has happened, as far as we can see in the score, is that everything has simply moved up five notes—but we have in fact travelled, in not much more than an instant, far from home. Then, just as quickly, we are back: the penultimate right-hand note of bar 2, unlike its counterpart in bar 1, reverses the direction of travel, leading to a *perfect cadence*—an unequivocal statement of tonal movement, here from dominant to tonic. The next time we hear the subject, it is firmly in the dominant key, whose arrival has been announced with a perfect cadence from bar 6 to bar 7. In order to see the symmetry here, we need to conceive of the various keys as a *circle* rather than a straight line. Indeed we should, because if we move from one key to another by a perfect fifth we will eventually arrive where we began (Figure 5.6). Thus the subject, and the piece based upon it, has been transformed by rotation.

Nor is rotational symmetry confined to tonality. In bar 7, the subject appears in the left hand first, then the right hand. The

FIGURE 5.5   Bach, *Inventio.* BWV 772.

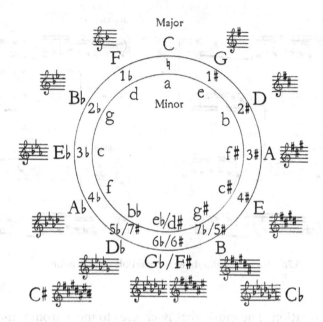

FIGURE 5.6   Circle/cycle of fifths.

point here is not simply that the order has been reversed between the hands, but that the subject now appears *above* its accompaniment, or the *countersubject*, not below as before. They have been *inverted*, not in the sense of melodic inversion as we saw in *Miserere nostri*, but an inversion of order. Bach's music, perhaps more than any other music we have, relies on and exploits this invertibility. Remember how the interval of a third above a note becomes the sixth below if inverted? Both intervals are consonant, and 'invertible counterpoint' depends upon such flexible intervals: thirds, sixths, and octaves (see Figure 5.7a). Fifths are tricky because they become dissonant fourths when inverted, but Bach knew how to incorporate fourths and other dissonant intervals into his invertible counterpoint so that the ear does not recognise them as dissonant.

A further symmetry exists here, one which combines melodic and tonal movement and depends upon both reflection

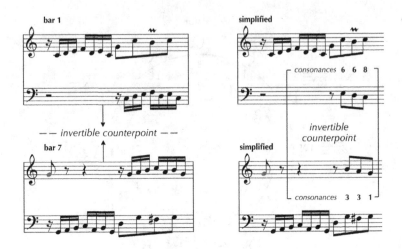

FIGURE 5.7a   Bach, *Inventio* showing invertible counterpoint.

and rotation. The music that Bach uses to move from tonic to dominant in bars 3 to 7, and which then takes us off on an extended journey around keys, is an inversion of the subject's opening: its first seven notes now fall then rise, instead of the other way around (Figure 5.7b). It has been transformed into a linear pattern of intervals, a free-wheeling 'sequence' of notes that repeat at regular intervals. Sequences such as this fulfil a purpose entirely different from that of the subject. Instead of defining a tonal area, they blur the boundaries. They thus

FIGURE 5.7b   Bach, *Inventio* showing reflection symmetry.

enable free movement from one key to another or provide a brief excursion away from an otherwise settled key or simply a sort of holding pattern. They need not have anything to do with the subject in terms of pitches and rhythms. Here, however, those basic constituents are exactly the same, except that here they have been inverted—or, in symmetrical terms, transformed by reflection. Furthermore, because sequences exist to move between keys—to make free but transitory associations until called to order by a cadence—and do so around the circle of fifths, they are constantly subject to rotational transformations [18].

All this from the first few bars of a piece in only two parts, which takes about a minute to play. And there is, of course, far more to say. We might, for example, note that Bach's fascination with invertible counterpoint persisted to the very end of his life, with the posthumously published collection *Die Kunst der Fuge*, an encyclopaedia of the different ways a single subject could be subjected to fugal and canonic treatment. Among the fugues are two 'mirror fugues', in which symmetrical reflection governs the music: each is really two fugues, but they are vertical mirror images, *rectus* and *inversus*, of each other. Another rotational symmetry is incorporated into the 'Goldberg' Variations, published in 1741: here every third variation is a canon, but the interval of the canon expands successively upward, eventually crossing over itself at the *canon at the octave* like a traveller crossing the International Date Line.

Small this Invention might be, but it tells us things that are of fundamental importance to our understanding of Bach's music. Invention and invertible counterpoint, both reliant on symmetrical transformations, are the foundation for all his often speculative music adventures. After Bach's death, the tonal ('diatonic') system, with tonic and dominant as its poles, continued to command the world of art music with its overarching mirror symmetry. In the diverse works of Haydn, Mozart, Beethoven,

Schumann (Clara and Robert), Mendelssohn (Fanny and Felix), and so many others, the tonal space defined by the tonic and dominant remained reassuringly closed, and the symmetrical underpinning of diatonic harmony continued to operate on a great variety of music.

As we have already seen, even when conventions and styles have changed radically, symmetry has not vanished from music. How could it? We would be wrong to think that 'total serialism' was the high watermark of musical symmetry. Indeed, at the very moment when that might have seemed the case, a group of musicians who were more interested in jazz and rock than high art and the academy were developing new symmetries in their own 'minimalist' compositions.

In Steve Reich's *Piano Phase* (1967), a repeated pattern is played on two pianos, one of which—imperceptibly at first—begins to fall out of step with the other, eventually becoming a quasi-independent voice in its own right, then moving gradually back into phase. Over the course of the piece, the same thing happens several times: as the pianos diverge their rhythmic interplay becomes frenetic, then calms as the second 'divergent' piano rotates to a point where it is playing different notes to the first, but in rhythmic unison—two notes sounding at the same time. The pattern of pitches never changes, but depending on the combination of notes being played, we intuit different rhythmic groupings; we hear accents emerge, then fade. There is a kind of counterpoint here, between the predictable and the unexpected—or rather the *unexpectable*, because our cognition of the patterns does not extend to predicting their effect.

Piano Phase and Bach's Invention both set their 'subjects' on a path and moving them forward through time. Symmetry plays an essential part in generating and governing that movement. Indeed, musical symmetries—whether they engender a sense of circularity or of propulsion, stillness, or activity—are perhaps even responsible for manipulating our experience of time itself.

EDITORS' COMMENTARY

Robert Quinney here offers a fascinating insight into how order and chaos appear in an entirely different context, music, and how they contribute to its aesthetic value. As he observes, without unharmonious dissonant intervals, music would be just one monotonous consonance after another!

Our chapter about planetary dynamics also examined the delicate bridge between order and chaos—Terquem's essay considered how chaotic planetary systems find themselves in synchronous orbits—and it is intriguing to observe that these ideas find parallels in music.

If we look at a Bach composition as it appears on the page, for example, we will observe a series of peaks and valleys. None of these shapes is completely regular: there is an element of unpredictability. But if we turn the piece into a series of regular shapes, the composition loses its magic and becomes monotonous. If we made it just a little bit more disordered, on the other hand, it would sound random and unappealing.

Quinney suggests that 'music ... is a fundamental constituent of the created universe: it is sounding evidence of the perfection of that creation', and this invites us to draw parallels with biology. In the following chapters, we will consider how biological organisms harness disorder and stochasticity in order to survive.

CHAPTER 6

The Interdependence of Order and Disorder: How Complexity Arises in the Living and the Inanimate Universe

Denis Noble

EDITORS' PREFACE

On the one hand, biological systems have developed mechanics to defend against randomness. For example, neurons need to figure out how to reduce signal noise when transmitting electrical impulses so that our body receives coherent instructions. But evidence is emerging that organisms harness randomness

DOI: 10.1201/9781003306986-6

in order to generate new functions to survive changes in their environments.

In this chapter, Denis Noble will consider specific examples of the harnessing of stochasticity, including McClintock's work on chromosome shuffling under radiative stress and the creation of new DNA in immune systems. The chapter finishes with a brief discussion on the implications of this in the philosophy of free will and in physics. We refer the reader to the work by Anant Parekh in the appendix, for a more technical description and an example of this harnessing of stochasticity works in calcium channels.

Can order and chaos be mutually dependent?

Standard biological theories of life see chance simply as enabling our genetic material, DNA, to accumulate random mutations. Natural selection then blindly filters the successful organisms from the unsuccessful ones in the struggle for existence. Order slowly emerges in this way during evolution. This is the essence of neo-Darwinist's modern synthesis. Amongst other consequences, it reduces living organisms, including us humans, to being mere vehicles for the transmission of DNA to future generations. Chance and order are independent processes. There is no way in which they can combine during the lifetime of the individual. Chance is experienced but not directly used by organisms themselves.

We must turn this view on its head. Contrary to the ideas of neo-Darwinism, living organisms harness chance as a way to solve the immediate problems they encounter in their environments. There are many ways in which this interdependence of order and disorder can occur.

Nearly a century ago a botanist in the United States, Barbara McClintock, was working on maize using a microscope to study the chromosomes in the cell nucleus. This is where the genetic material, DNA, is kept. She found that under environmental stress, such as radiation, the plants start shuffling large parts of their chromosomes. They were using chance (the shuffling) to find combinations that would better enable them to survive. She published this ground-breaking discovery in 1953 in the journal *Genetics*, only to find that no one took any notice. Not a single scientist even asked for a copy of the paper. Very disappointed, she stopped sending her work to *Genetics*. Thirty years later, in 1983, after other scientists had also found such large 'jumping genes', she was awarded a Nobel Prize for her discovery—at the age of 81.

Even the award of a Nobel Prize, though, did not change the paradigm. So, is such harnessing of chance in organisms just a rare exception?

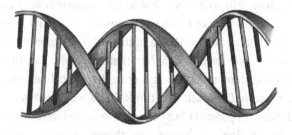

Well, no. I have to tell you that it is far from an exception. You are harnessing chance in this kind of 'dance between order and chaos' all the time. All organisms with immune systems do so. When they encounter a new virus, bacterium, or any other foreign body, they tell a very specific part of their genomes, the part that could make a new antibody to tackle the invader, to produce literally millions of new DNA sequences. Your immune system then works out which random variations work in neutralising the invader. It then tells the immune system cells that succeed to reproduce. That is how you fight off the new infection. In this way, chance is used to produce something, the immune response, which is very far from chance. You, in effect, *create* new DNA sequences and you do that even during your own lifetime. You don't have to wait for evolution to do so in your descendants.

You are not conscious of your immune system doing this. But could a conscious choice be working in much the same way?

Could your sense of free agency derive from the interdependence of order and disorder?

Reductionists would say no. You only think you have a real choice in how you behave. That feeling is just an illusion. Your genes and your neurones (created by your genes) make you do what you decide to do.

Again, I disagree. Random disorder occurs at all levels in your nervous system. Individual proteins forming channels in the neurones are opening and closing, partly at random, all the time. I think we use this and other forms of randomness in our nervous systems to 'spin the wheel of chance' just as the immune system 'spins the wheel of chance' in your DNA. Just as the immune system generates an unlimited number of DNA variations, your nervous system can harness the chance to develop an unlimited repertoire of behaviour. The neural processes that generate that repertoire can then mesh with your social interactions to generate the 'logical' response to the situation you find yourself in. I think this is why creative activity is unpredictable in prospect but can be rationalised in retrospect.

The harnessing of chance can therefore be the means for resolving the tension between micro-level and macro-level explanations of what you decide to do. Free agency in this view is relatively independent of micro-level causes but not independent of macro-level causes. Nor would you want it to be. If you are like me, you will be quite happy to be determined in your actions by what you see as the right fit to what the situation demands, just as your immune system finds the right fit for the invading virus. The difference is that you are aware of your neural actions. You are not aware of your immune system's actions.

These are the processes by which living organisms can be seen through the interdependence of order and disorder. But what about the inanimate universe?

Imagine a universe in which all matter is completely uniformly distributed. Perhaps it was like that when the universe was forming from the 'Big Bang' over 13 billion years ago. Clearly this is not the universe we now know. It is full of strange objects: circles, spirals, horseshoes, vast dust clouds of various shapes and sizes, and of course even stranger objects in modern theories of physics, such as black holes, dark matter, and dark energy. And the most wondrous objects of all: living organisms!

How did such complexity arise from what may have been uniformity and from a perhaps infinitesimally small beginning?

One answer to the origin of complexity is surprisingly simple. There are forces between objects. Some forces, like gravity and the opposite poles of magnets, attract while others, like the positive poles of two magnets, repel. This is to find symmetry in modern science that resembles the symmetry of yin and yang in ancient science.

Because of these forces between polar opposites, if we could turn the clock back and distribute the matter of the universe evenly, the particles would immediately start their dance of attraction and repulsion. As they do so, they inevitably form networks of interactions. No particle would initially be in a privileged position, but as they attract each other they would congregate to form clumps. Once that happens, we break the symmetry of a perfectly uniform universe. Those clumps would form initially as clouds and then as stars and planets.

Breaking symmetry in an unstable system is easy. Small chance events can do the trick. Imagine a ball placed exactly at the top of a hill with a shallow enough top for the ball to have the possibility of staying put. It might stay there indefinitely if there were no chance of perturbations. But the slightest wind would displace it from the peak, and it would start to roll downhill all the way if it encounters no insuperable obstacles. Depending on how fine the slope is at the top it may initially move extremely slowly, but then more rapidly as it experiences steeper slopes. On the way that rolling ball may trigger many other events, such as landslides, that may in turn kill unsuspecting climbers, in turn disturbing their family and friendship structures … the list is endless. Once symmetry has broken, further events can occur simply because of the energy and matter gradients that form, and the chance encounter with other events and situations. There is a continual 'becoming', described by some oriental philosophers as conditioned arising. These are processes with ever more possibilities arising because each arising forms the conditions for many others.

We can see this kind of process at work in the weather systems of the sky above us. On a perfectly clear day, the sky looks uniform as the apparently evenly distributed particles in the atmosphere scatter sunlight to form the uniform blue colour we see. But the stillness and uniformity mislead us. All the time the atmosphere is exchanging heat, water, and gases with the oceans and continents. Convection currents arise as warmer parts rise and colder parts fall. Water particles accumulate. The interactions between them form the wide diversity of cloud structures, from relatively simple smooth planes, to the fiendish complexity of a tornado. No one thing 'makes' the tornado. It makes itself. The forces of matter under the right conditions ensure that these structures should develop. As the complexity increases so does movement within them. They spin rapidly and rhythmically like tops. The movement in the strongest tornados is so strong that they create immense damage as they hit land and dissipate their energy in a frenzy of destruction.

From a distance above the earth, they appear as spirals. So do spiral galaxies in the depths of space. These also rotate. Our whole galaxy, the one that forms the Milky Way, is rotating, so we are rotating with it, probably around a black hole at the centre of the galaxy. There are rotating structures everywhere in space as well as on the earth.

We don't need any abstruse kind of theory to explain these formations, both celestial and cosmic. The equations of Newtonian motion suffice, although relativistic effects must also be involved. A similar process creates these rotating structures in both cases. We call them and many other self-sustaining structures of networks 'attractors'. The system tends towards these attractors, which explains the name. Note also that no particular part of the network is the cause of the attractor. The spiral and its circular motion are properties of the whole network of interactions. They attract more matter and energy to themselves. They are states of the network that attract other parts of the network until the whole network dances to the tune of the attractor.

COMMENTARY

Neo-Darwinism maintains that variation among organisms is random, but not functional in its own right. Randomness serves no purpose in prolonging the survival of a set of organisms and is very much a 'passive' part of natural selection. In other words, a combination of idiosyncratic events causes variation, and naturally advantageous mutations stick around. It is the idea that genetic variation is random and not actively harnessed for purpose.

Noble gives this idea a name, 'blind stochasticity', and he references examples where randomness is actively harnessed instead of acting passively. Under some environmental challenges, organisms can accelerate mutation rates in variable parts of the genome, as if some targeted process is going on to speed up the rate of the genetic slot machine. He gives two examples. Firstly, in the immune system, the mutation rate in the variable part of the genome is accelerated in response to a new antigen. Secondly, in bacteria, there is a five-order-of-magnitude acceleration of reorganisations in the genome to help cope with this additional environmental stress. An example of this accelerated mutation is discussed in [19] (see 'References'), where the authors have studied more vigorous mutation rates in response to starvation. The most referenced phenomenon is the FC40 strain of Escherichia coli. Experiments have been set up where E. coli are bred not to be able to use lactose, with episomes containing the lactose gene placed in the bacteria. When exposed to agar plates where the only carbon source is lactose, mutant lac+ colonies arise rapidly. This harnessing of mutation is similar to 'explore' states instead of an 'exploit' states in the field of reinforcement learning today.

This is not the only link between stochasticity and biology. In particular, Noble has previously referenced [20] a study where the random distribution of a cell population seems to be retained in all parts of the distribution. Chang et al. [21] showed that cells cherry-picked from extreme points in a distribution retain their 'memory' of the original distribution. This phenomenon is intriguing, as you would expect that bacteria sampled on the extreme ends of a distribution would have no information about whether it was bimodal or unimodal.

CHAPTER 7

A Philosopher's Perspective on the Harnessing of Stochasticity

Sir Anthony Kenny

EDITORS' PREFACE

There are many interesting points that can be made about the philosophy of stochasticity in biology and its bearing on the debate concerning free will. One philosophical matter is the difference between systems that are practically random and systems that are truly random. Another matter is the implications of stochasticity in biology on the free will debate, as touched on briefly by Denis Noble.

These topics will be discussed in more detail in the following essay by Sir Anthony Kenny. Kenny examines some of the philosophical issues that Noble writes about in his essay. Specifically, he argues against the idea that the reasoned actions of human beings

DOI: 10.1201/9781003306986-7

themselves exploit stochasticity at lower levels and encourages the separation of the psychological and the physiological. He relates Noble's ideas to the concepts of upward and downward causation and the ideas of constraint and control.

In Denis Noble's fascinating paper, there were many interesting points deserving a response. I wish to focus on just two of them: the harnessing of chance and the possibility of free agency. I find his discussion of the first point admirable, but I have reservations about his treatment of the second point.

Living organisms, Noble tells us, harness chance in order to solve the problems they encounter in the environment. The chance he has in mind is not pure randomness but rather stochasticity, that is to say, unpredictability. Stochasticity, unlike randomness, is an epistemological rather than a metaphysical notion: it is primarily concerned with what we can know, rather than what is the case at a fundamental level. If we are told something is unpredictable we can ask: unpredictable by what system of prediction? (If something is random it cannot be predicted by any system whatever.) Molecular events are unpredictable by molecular physics, but they may be predictable on the basis of some other system. As Noble points out that an event that is unpredictable at one level—say the molecular—may be predictable at another, higher, level—say the cellular. That opens up the possibility that if so, then in addition to bottom-up causation (the lower level causing events at the upper level) we have top-down causation (the upper level causing events at the lower level). So far, I find Noble's account highly illuminating.

However, I am not so persuaded when Noble applies his model to the reasoned actions of human beings. It is tempting to think of human choice as being the top level of causation and to suggest that like physiology it exploits stochasticity at lower levels. I believe that this temptation should be resisted. A choice of action is not a form of causation, not even of top-down causation, and the psychological level is not above the physiological level but operates in parallel with it.

Of any tract of human life, there are always two narratives to be told: the physiological and the psychological. The narratives are expressed in different languages, and the first step to understanding their relationship is to separate each language from the other without describing either of them in the other's vocabulary. In describing the physiological operation of the immune system, Noble says, 'your immune system works out which random variations work in neutralising the invader. It then tells the immune system cells that succeed to reproduce'. The expressions 'work out' and 'tell' belong to the psychological, not the physiological vocabulary. I would like to see these metaphorical terms replaced by expressions suitable for physiological causation.

Noble is undoubtedly right to reject the reductionist claim that our feeling of choice is an illusion. Many of our actions are free and undetermined. When I perform an action I enjoy freedom, provided only that I have both the ability and the opportunity to do something else instead. It is possible to discover whether I possess this ability and this opportunity without knowing anything about the physiological processes occurring in my body at the relevant time.

Suppose that I am sitting in an armchair. I look at my bookcase with its 500 books, get up and take down the poems of John Donne to check a quotation. In the course of this, there will be causal processes linking my brain, central nervous system, muscles, and so on. What is the relation between my action and these causal processes? Shall we say that the two narratives report the same events in different languages? Not so, because the way in which we identify and individuate an event may be different in different languages.

Noble offers a parallel between my fetching the book and the creation of an antibody by the immune system. There can indeed be a similarity between the operation of the immune system and some conscious human choices, such as the choice of words in a sentence, and the choice of sentences to express a thought. Several different formulations may come into one's head before one chooses the right one and sets out to type it. But the similarity is far from being the exact parallel that Noble supposes.

'The immune system', he says, 'generates an unlimited number of DNA variations, your nervous system can harness chance to develop an unlimited repertoire of behaviour'. But a repertoire is not a set of actualities, but of possibilities. Besides my actual behaviour of taking down the Donne book, there are 499 possible behaviours of taking down one of the other books. Possible behaviours, unlike antibodies, are not empirically detectable entities. There is no reason to think that they exist in actuality either psychologically or physiologically, either as conscious thoughts or as corporeal states.

Noble is right to reject the idea that human actions are subject to determinism. But the verb 'determine' is ambiguous: it may mean 'constrain' or 'control'. Upward causation is constraining but not controlling: an example is physical causation. The laws of physics determine what I do by setting boundaries to my abilities: I cannot do anything that is physically impossible. But the laws do not control what I do: they do not settle which particular action I perform. The distinction between constraining and controlling may be illustrated by a comparison with the game of chess: the rules of the game constrain what moves are possible and the players control what form an individual game takes.

Downward causation, unlike upward causation, may be controlling no less than constraining, as for instance when a cell reconfigures a DNA sequence. It is here that the notion of harnessed stochasticity is most appropriate. Earlier I objected to the use of metaphorical psychological terms such as 'work out' in the description of physiological processes. However, the metaphor 'harness' is unobjectionable because it fits both kinds of processes. It brings out that stochasticity at a lower level is actually a necessary condition for teleology at an upper level. If there was determinism at the molecular level, there would be nothing for any upper level to harness—in terms of the metaphor, there would only be a runaway horse. Thus we see that teleology and stochasticity are not opposed but rather complementary.

At the upper levels of the biological ladder, we find not only top-down *harnessing* of stochasticity but also top-down *creation* of stochasticity. This already emerges at a level lower than the humans, in species that indulge in play. Play is stochasticity that by definition is unharnessed, that is to say, activity that is unpredictable and that serves no immediate purpose. Overall, of course, it serves a developmental purpose, since it enables an animal to experiment with different methods of achieving goals, and this in its turn is an evolutionary advantage.

At the human level play becomes organised into games, where stochasticity is not so much harnessed as confined—whether solely by rules, as in chess, or also physically as in billiards. In all but the most boring of games, the actual moves or ploys are not predictable from either the rules or the constraints.

The relation then between the psychological and the physiological narrative of my taking the book from the shelf can be simply stated. The physiological story constrains the psychological one: all kinds of physiological states and processes are necessary conditions if I am to be able to perform the action of grasping the volume of Donne. But the physiological story does not control the psychological story: no physiological state is a sufficient condition of the free choice that decides which volume is selected.

COMMENTARY

Denis Noble: Tony Kenny and I have been debating these issues for over 50 years now, since we became young tutorial fellows at Balliol in the early 1960s. We have both come a long way since then! I think the idea of harnessing of stochasticity has changed the terms and context of the debate. I agree that there are important issues still to be ironed out, which are beyond the immediate scope of this project. Suffice it to say that, in some sense, our rational choices do have physiological consequences. The choice of a professional runner to exercise regularly demonstrably changes molecular details in his body at the level of nucleic acids and proteins, for example. I suspect the full debate between us on the nature of that causation could be the subject of a whole book that perhaps Kenny and I may write!

CHAPTER 8

Postscript

A Dialogue between Denis Noble and Benedict Rattigan

Noble: I am amazed at the way this has all panned out, Red. I had no idea, when you and I first met a year or two ago, where this was heading. You've led me into an area that I have never explored before, but it's obvious to me now that it's where I should have been looking.

Rattigan: The order–disorder dichotomy is a rather unexpected symmetry, isn't it?

Noble: Yes, very much so. People tend to think there's either order or disorder, but in fact you need *both*, and in very many different ways. And what's happened just recently is this insight is turning out to have huge practical importance, too.

Rattigan: In what way?

Noble: Over the last 50 years or so, the developed nations of the world have plugged almost $300 billion into the so-called war on cancer. And what do we find? With third- and fourth-stage cancers, life expectancy has

DOI: 10.1201/9781003306986-8

hardly changed at all. Now, why could that be the case? With late-stage cancers, we aggressively attack the cancerous growth by zapping it with radiotherapy or chemotherapy. What happens? The cancerous cells realise they're being attacked, and they start randomly accumulating different forms of themselves to see whether they can find a solution. The cancerous tissue evolves, in other words, precisely because we're attacking it: it 'harnesses' disorder to develop new ways of surviving. Now, I was brought up on the central dogma of molecular biology on the idea that genes make us what we are, and there's a kind of determinate process from the DNA through to the proteins, through to us. But it now seems that all of that, to put it bluntly, is close to being a big mistake, because it fails to take account of the fact that the randomness gets used. And that's the key. I've just taken part in a conference in Boston in which we explored the question, 'Could the harnessing of stochasticity be relevant to cancer?' Now, this idea has penetrated already the American Association for Cancer Research, the world's largest organisation of researchers on cancer. Sometimes, as a scientist, it's through discussion and interaction with people that things become clear, and that is the case here. They believe what we're onto might be extremely important, and I'm part of the Cancer Evolution Working Group they've set up to explore it.

Rattigan: And it's such a simple idea, isn't it?

Noble: It is simple, but it's also very profound. The more I've thought this through, the more it seems that is the fundamental symmetry of the universe. It must have been the first symmetry to be broken—if we indeed believe the so-called Big Bang model, the development of the universe—to get from what might have been a fairly uniform beginning to the huge range of structures that we now find in the universe. Somewhere there must have been that breaking down into disorder. But that's continued, it seems to me, all the way through to the origin of life, and to the development of life, and the way life itself has evolved. It's a continuous process of using that symmetry of order and disorder. And if that is so, then symmetry itself *must be* the fundamental principle of Nature. It has to be so.

Rattigan: That's a big claim, but I think you're right: all other laws are secondary to the way in which the symmetry between order and disorder occurs.

Noble: There's a wealth of evidence to support this. Take physics, for example. The whole development of thermodynamics is based on this idea. At large numbers of elements, the jiggling around or stochasticity becomes uniform. And the question of the breaking of symmetry has, of course, become a key feature of the way in which the standard model of the universe has been developed.

Rattigan: And it seems to apply everywhere, doesn't it?

Noble: I think that's right. I was fascinated by the contribution of Robert Quinney on the role of the order–disorder principle in music. And the same seems to apply to logic. I mean, the mind boggles at what the logician was doing! Joel David Hamkins builds logic on logic on logic on logic out of the whole principle! That, too, is a gem of a piece. Of course, we've known about the whole issue of symmetry for a long time, but what we've not seen before is how fundamental and far-reaching it is, nor how paradoxical it can be.

Rattigan: The inconceivable nature of Nature!

Noble: Hah, precisely!

Rattigan: I'm intrigued by Professor Hamkins's idea that there are many transformations of the idea of symmetry. One of the questions raised by The Language of

Symmetry is this: how can life's ordering principle—a law which must be all around us, in everything we touch and see—remain undiscovered after all these centuries? And the answer, I believe, may lie here, in this idea of transformation.

Noble: Go on.

Rattigan: In their quest for the ordering principle, scientists and philosophers have been trying to identify a law that they assume must be fixed and objective, like Archimedes' principle or Newton's laws of motion. But quantum mechanics suggests that, at the deepest level, the universe is paradoxical and subjective. For example, we've learnt that light can present itself in contrasting ways. To some observers it takes the form of a particle, which is localised, whilst to others it appears as a wave, which is spread out. Yet whilst these two interpretations, particle and wave, appear to be contradictory, both are true.

Noble: Indeed.

Rattigan: Symmetry, however, has an infinitude of different faces. Having no deeper cause than itself, it ceaselessly mutates in a sequence of oppositional symmetries: sometimes it is orderly, and sometimes chaotic; often it is logical, whilst at other times it's contradictory; it is always ubiquitous, yet often quite difficult to see. It can express itself as harmony and opposition simultaneously, or as transformations that leave an object unchanged. It is deeply symmetrical. Which of its complementary aspects are observed depends entirely on the observer, however. Mesopotamian mythology describes it in terms of harmony, whereas the Ancient Egyptians saw it as a law of opposition and conflict. Both of these subjective interpretations of symmetry are valid, but they contradict each other. And it is here, with the crucial polarity of the observer and the observed, that quantum mechanics and relativity coincide with symmetry. In fact, I would suggest that without an observer to give it form, we might even wonder what symmetry, in itself, really 'is'!

Noble: I think that perspective is very interesting. I've often asked myself the question, how did we manage for a hundred years after the discovery of quantum mechanics to go on pretending nothing had happened? We've lived with the quantum model at a very deep level for over a century now, but it seems to have had very little impact on the way we understand the macroscopic world.

Rattigan: And yet it has applications at every level of life.

Noble: I think that's right.

Rattigan: From a philosopher's perspective, it seems clear that the time has come to discard the objective Newtonian mindset. Symmetry isn't just an expression of orderly transformation, it's far more subjective and fluid than that. And that's why the ordering principle has so long eluded physicists—it has never occurred to anyone that it might be a shape-shifter, a law that expresses itself in contradictory ways under different circumstances.

Noble: It's certainly true that symmetry has always been seen by scientists as an expression of invariance or order, and for this reason, it's assumed that it cannot be the guiding principle of a universe which is both orderly and chaotic. The notion that order depends upon chaos or disorder goes against everything we're taught.

Rattigan: And yet we've demonstrated that the order–disorder dichotomy is important in several different fields.

Noble: Yes, we have. I didn't initially see it as a symmetry issue, but it evidently is. In fact, it's clear to me now that it is the most fundamental symmetry there is!

Rattigan: The bit that scientists have routinely ignored, the point where order and chaos meet—

Noble: That's where it all happens! This simple insight is the sort of break-through we academics dream of. I've already mentioned cancer, but there are several other areas of great importance. We depend on antibiotics, for example, but what have we done? We've used them so widely in agriculture and many other fields that we have provoked the bacteria themselves to harness stochasticity to escape the antibiotics. We now have to find ways of stopping the process by which we ourselves are triggering the evolution of the bacteria.

Rattigan: You know, it's curiously symbiotic that you and I have each spent several years researching the same concept, the relationship between order and chaos, but approaching it from opposite directions. My interest has been in ancient philosophy and mythology, whereas you are a scientist. I have developed the general overview, and you the essential detail. Your idea of the harnessing of stochasticity has applications in many fields, whilst it is evident that its elegantly simple ordering principle, symmetry, has infinite reach and is hiding in plain sight all around us.

Noble: I think that's right. Nature likes simplicity, but it also likes to hide. It communicates with us through contradiction and paradox, but it is a universal language.

Rattigan: And if we learn to speak the language of symmetry—

Noble: Then new worlds could be ours to discover.

A Response to Professor Noble's Paper: Ordered Disorder to Drive Physiology

Anant Parekh FRS—Professor of Physiology with Frederick B. Parekh-Glitsch and Daniel Balowski

ION CHANNELS OCCUPY A central and indispensable role in the physiology of animal cells. These remarkable miniscule machines move ions rapidly in and out of cells, changing the electric potential across the cell surface. Ion channels might be considered the quarks or building blocks of the nervous system; their activity forms the core of our ability to detect and respond to our environment, form thoughts, and translate them into actions. Ion channels are found in all cells, across all species where they are essential for the ability of the immune system to combat infection and for the heart to beat.

Ion channels are unique amongst molecules that populate the physical and life sciences because the proteins can be studied at an individual level using a form of electrophysiology called patch clamp recording. Physicists and chemists cannot tell us how one molecule will behave; they can only predict the behaviour of the population as a whole. Studies on single ion channels have revealed that channels often exist in two main states: open and closed (Figure 9.1). Open channels are the functional form, permitting ions to move rapidly across the cell surface. Surprisingly, detailed single-molecule studies have shown that many voltage-gated Ca^{2+} channels, which drive neurotransmitter release or the ability of the heart to pump blood, have a low likelihood of opening (or open probability) upon stimulation. The open probability is typically <0.1. This means that, if stimulated ten times, the

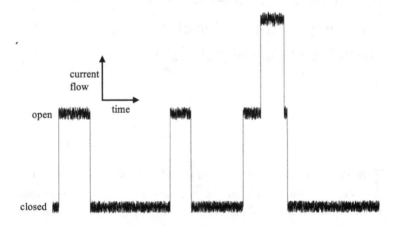

FIGURE 9.1 Single channel properties of a K^+ channel. Closed denotes the channel in the closed (non-conducting) state and open represents the functional, conducting conformation. The channel is in the closed state for ~70% of the time shown. Note the two open events that superimpose towards the end of the trace, indicating the presence of two independent channels. Openings typically last 1 millisecond.

Ca^{2+} channel will open, on average, just once. This is hardly an effective way to rapidly propagate electrical signals across a nerve cell, where speed and timing are critical. How has biology managed to extract order from a stochastic system? Perhaps the first solution came by ensuring a favourable statistical outcome by increasing numbers. Cells often express hundreds or thousands of the same type of Ca^{2+} channel at the same time. Consider a cell with 1,000 channels, not an unrealistic number in heart muscle for example. Each individual channel has an open probability of say 0.01. But, with 1,000 channels, 10 channels on average would open with each stimulus. We do not know which 10 channels will open, but 10 will open. Since each channel can conduct many Ca^{2+} ions, ~10 open channels at the same time will lead to a significant change in electric potential across the cell, ensuring the propagation of electrical activity. In this regard, for an electrical signalling mechanism, it matters not which ion channels open as long as a few do to change the potential across the cell surface.

Electric signalling is harnessed in relative simple metazoans and multicellular organisms ostensibly to allow the animal to move rapidly towards a food source or away from a predator. As biological complexity increases, more nuanced responses are required as well as stimulus-induced long-lasting changes in the complement of genes expressed. Most of these responses are driven by Ca^{2+} channels, which double up as signal transducers providing trigger Ca^{2+} to activate downstream pathways independent of electrical activity. How does the Ca^{2+} channel turn on the relevant pathway, such as a protein (called a transcription factor) that moves to the nucleus to control the expression of a particular set of genes? If the transcription factor is located away from the channel, it would not know precisely when the channel opened and this would induce a further level of unpredictability in addition to the uncertainty in knowing when a channel will open.

It turns out that Ca^{2+} channels have private conversations with the transduction pathway, accomplished through a physical interaction between the channel and the transcription factor that relays information from the channel to the cell nucleus, where most of the genetic material is housed. Typically, a triumvirate of additional proteins is required to ensure these restricted communications take place with functional impact on the cell: a scaffolding protein that binds to the channel, the transcription factor itself, and a protein that ensures the factor is activated only when the appropriate stimulus is received.

Producing three proteins for every Ca^{2+} channel is energetically costly for a cell and imposes physical constraints on the cell surface. Moreover, Ca^{2+} channels can signal to different transcription factors and so different sets of triumvirates are required and all cannot be juxtaposed against each Ca^{2+} channel. How does the cell ensure signal transduction through different pathways is faithfully recruited by Ca^{2+} channels that behave stochastically? Evolution has employed the process of channel clustering to solve this problem. In many cell types, Ca^{2+} channels are concentrated into small regions, often accommodating tens of channels closely spaced together. Each channel can bind a different triumvirate but, because the channels are so packed, the opening of any one channel will generate a local Ca^{2+} signal that is detected by all the triumvirates. Which channel in a cluster opens is therefore not important; it is sufficient for any one channel to activate. In this way, biology has harnessed the principles of carefully selecting neighbours and habiting in a dense local population to convert the inherent disorder at the level of a single molecule into a high fidelity and predictable outcome. At a microscopic scale, this is an example of how collectivism triumphs over individualism.

Perhaps biology has managed to reconcile a form of quantum mechanics with Newtonian pre-determinism; figuring out a way to convert the stochastic and unpredictable behaviour at a single-molecule level into a smooth, reliable, and predictable macroscopic response.

References

1. Gell-Mann, M., *Phys. L&T.* 8, 214 (1964).
2. Zweig, G., "An SU3 Model for Strong Interaction Symmetry and its Breaking," CERN Report No. TH 412 (Geneva, 1964) and "Fractionally Charged Particles and Sub," in Symmetries in *Elementary Particle Physics* (Academic Press, New York, 1965), p. 192.
3. Friedman, J.I., Kendall, H.W., *Ann. Rev. Nucl. Sci.* 22, 203 (1972).
4. Taylor, R.E., *Rev. Mod. Phys.* 63, 573 (1991).
5. Friedman, J.I., *Rev. Mod. Phys.* 63, 615 (1991).
6. Wu, C. S., Ambler, E., Hayward, R. W., Hoppes, D. D., Hudson, R. P., Experimental Test of Parity Conservation in Beta Decay, *Phys. Rev.* 105, 1413 (1957).
7. Clausius, R., On the Application of the Mechanical Theory of Heat to the Steam-Engine, as found in: Clausius, R. (1865). *The Mechanical Theory of Heat – With its Applications to the Steam Engine and to Physical Properties of Bodies.* London: John van Voorst, 1 Paternoster Row. MDCCCLXVII.
8. Müller, I., *A History of Thermodynamics. The Doctrine of Energy and Entropy.* (Springer, Berlin, 2007), p. 134.
9. Landau, L.D., Lifshitz, E.M., *Statistical Physics.* Course of Theoretical Physics (Vol. 5, 3rd ed.). Oxford: Pergamon Press. ISBN 0-7506-3372-7.
10. Hawking, S. *A Brief History of Time is Based on the Scientific Paper.* Hartle, J.B., Hawking, S.W., Wave Function of the Universe, *Phys. Rev.*, 28, 2960 (1983).
11. Joel David Hamkins. "Every group has a terminating transfinite automorphism tower". *Proceedings of the American Mathematical Society* 126.11 (1998), pp. 3223–3226. ISSN: 0002-9939. http://doi.org/10.1090/S0002-9939-98-04797-2. arXiv:math/9808014 [math.LO]. http://jdh.hamkins.org/everygroup/.

12. Simon Thomas. "The automorphism tower problem". *Proceedings of the American Mathematical Society* 95 (1985), pp. 166–168.

13. Scolica enchiriadis (ca. 850 ce), trans. Lawrence Rosenwald, in Pierro Weiss and Richard Taruskin, *Music in the Western World: A History in Documents* (Belmont, CA: Thomson/Schirmer, 2007), 34.

14. The distinction, first made by Richard Taruskin in a record review of 1987, later became the title of his collection of essays, *Text and Act: Essays on Music and Performance* (New York: Oxford UP, 1995).

15. For one example of non-Western musical symmetry, see Dave Benson, *Music: A Mathematical Offering* (Cambridge: Cambridge University Press, 2006), 322–324.

16. Josef Hauer (1883–1959) had, in fact, experimented with a 'twelve-tone' technique a few years before Schoenberg, who saw some of Hauer's keyboard music in 1916. Other composers had also 'discovered' the technique. See Taruskin, *Oxford History of Western Music, Vol. 4: Music in the Early Twentieth Century* (Oxford: Oxford University Press, 2010), 680–686.

17. Joan Peyser, *To Boulez and Beyond* (revised edition) (Lanham, MD: Scarecrow Press, 2008), 147. The quotation first appears in the same author's *Pierre Boulez: composer, conductor, enigma* (1977), which reported conversations between the author and Boulez during his tenure as Music Director of the New York Philharmonic Orchestra.

18. For a full analysis of BWV 772, see Laurence Dreyfus, *Bach and the Patterns of Invention* (Cambridge, MA and London: Harvard University Press, 1996), 10–26.

19. Bridges BA. 1997. Hypermutation under stress. *Nature* 387, 557–568. http://doi.org/10.1038/42370.

20. Noble D. 2017. Evolution viewed from physics, physiology and medicine. *Interface Focus* 7, 20160159. http://doi.org/10.1098/rsfs.2016.0159.

21. Chang HH, Hemberg M, Barahona M, Ingber DE, Huang S. 2008. Transcriptome-wide noise controls lineage choice in mammalian progenitor cells. *Nature* 453, 544–547. http://doi.org/10.1038/nature06965.

Index

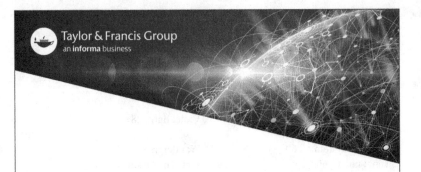

Printed in the United States
by Baker & Taylor Publisher Services

Printed in the United States
by Baker & Taylor Publisher Services